I0064409

Plant Biotechnology

Plant Biotechnology

Edited by
Jason Angstrom

Larsen & Keller
www.larsen-keller.com

Plant Biotechnology
Edited by Jason Angstrom
ISBN: 978-1-63549-647-5 (Hardback)

© 2018 Larsen & Keller

⊟ Larsen & Keller

Published by Larsen and Keller Education,
5 Penn Plaza,
19th Floor,
New York, NY 10001, USA

Cataloging-in-Publication Data

Plant biotechnology / edited by Jason Angstrom.
    p. cm.
Includes bibliographical references and index.
ISBN 978-1-63549-647-5
1. Plant biotechnology. 2. Agricultural biotechnology.
I. Angstrom, Jason.
SB106.B56 P53 2018
631.523 3--dc23

This book contains information obtained from authentic and highly regarded sources. All chapters are published with permission under the Creative Commons Attribution Share Alike License or equivalent. A wide variety of references are listed. Permissions and sources are indicated; for detailed attributions, please refer to the permissions page. Reasonable efforts have been made to publish reliable data and information, but the authors, editors and publisher cannot assume any responsibility for the vailidity of all materials or the consequences of their use.

Trademark Notice: All trademarks used herein are the property of their respective owners. The use of any trademark in this text does not vest in the author or publisher any trademark ownership rights in such trademarks, nor does the use of such trademarks imply any affiliation with or endorsement of this book by such owners.

For more information regarding Larsen and Keller Education and its products, please visit the publisher's website www.larsen-keller.com

Table of Contents

Preface

Plant biotechnology is the study and practice of breeding plants in order to make them profitable. The process involves giving better characteristics to plants by various methods like molecular markers, reverse breeding, doubled haploids, genetic modification, pharming, etc. This book elucidates the concepts and innovative models around prospective developments with respect to plant biotechnology. Most of the topics introduced in it cover new techniques and the applications of the subject. It explores all the important aspects of the area in the present-day scenario. This textbook is an essential guide for both academicians and those who wish to pursue this discipline further.

To facilitate a deeper understanding of the contents of this book a short introduction of every chapter is written below:

Chapter 1- Men have cultivated selective plants to use them for food, shelter and medicine through history. To simplify and boost cultivation, the plants are isolated from their source and cultured separately under a controlled environment. A popular method, plant tissue culture has been used for the purpose. This is an introductory chapter which will introduce briefly all the significant aspects of plant biotechnology and plant tissue culture.

Chapter 2- With many plants not being able to produce seeds or lacking the ability to go through vegetative reproduction, techniques such as micropropagation is used to produce huge amount of plantlets. To heed to the need of producing large-scale seeds, artificial seed production is utilized. This chapter discusses the methods of micropropagation and artificial seed production in a critical manner providing key analysis to the subject matter.

Chapter 3- The aggregation found in cells while suspended in a liquid medium is known as cell suspension culture. The common manner of cell suspension is by transferring a friable callus which helps in the dispersing of cells and the eliminating of large callus pieces. The topics discussed in the chapter are of great importance to broaden the existing knowledge on cell suspension cultures.

Chapter 4- Plant genetic engineering helps in introducing new traits in plants by modifying their DNA. The methods used in this process are gene gun bombardment, microinjection, agrobacterium, electroporation etc. The chapter strategically encompasses and incorporates the major components and key concepts of plant genetic engineering, providing a complete understanding.

I would like to share the credit of this book with my editorial team who worked tirelessly on this book. I owe the completion of this book to the never-ending support of my family, who supported me throughout the project.

Editor

Basics of Plant Tissue Culture

Men have cultivated selective plants to use them for food, shelter and medicine through history. To simplify and boost cultivation, the plants are isolated from their source and cultured separately under a controlled environment. A popular method, plant tissue culture has been used for the purpose. This is an introductory chapter which will introduce briefly all the significant aspects of plant biotechnology and plant tissue culture.

Plant Tissue Culture

For thousands of years man has been dependent on plants for food, shelter, medicine and many other purposes. Throughout history he has cultivated and selected the useful plants more suited to his needs. The latest advances in plant biotechnology provide potential to make improvements much more quickly than by conventional plant breeding. Studies on all aspects of plant development and multiplication in whole plants are often complicated by interactions between the various processes that underlie growth and development. It is, therefore, desirable to simplify matters so that controlling influences can be easily identified and studied. This can be done by isolating and culturing parts of the plants in vitro. Plant tissue culture has become popular among horticulturists, plant breeders and industrialists because of its varied practical applications. It is also being applied to study basic aspects of plant growth and development. The discovery of the first cytokinin (kinetin) is based on plant tissue culture research.

The earliest application of plant tissue culture was to rescue hybrid embryos and the technique became a routine aid with plant breeders to raise rare hybrids, which normally failed due to post-zygotic sexual incompatibility. Currently, the most popular commercial application of plant tissue culture is clonal propagation of disease-free plants. *In vitro* clonal propagation, popularly called micropropagation, offers many advantages over the conventional methods of vegetative propagation: (1) many species (e.g. palms, papaya) which are not amenable to *in vivo* vegetative propagation are being multiplied in tissue cultures, (2) the rate of multiplication *in vitro* is extremely rapid and can continue round the year, independent of the season. Thus, over a million plants can be produced in a year starting from a small piece of tissue. The enhanced rate of multiplication can considerably reduce the period between the selection of a plus tree and raising enough planting material for field trials. In tissue cultures, propagation occurs under disease and pest-free conditions.

An important contribution made through tissue culture is the revelation of the unique capacity of plant cells, called "cellular totipotency". It means that all living plant cells are capable of regenerating whole plants irrespective of their nature of differentiation and ploidy level. Tissue culture also provides the best means to elicit the cellular totipotency of plant cells and, therefore, it forms the backbone of the modern approach to crop improvement by genetic engineering. Regeneration of plants from cultured cells has also found many other applications. Plant regeneration from cultured cells is proving to be a rich source of genetic variability, called "somaclonal variation". Several somaclones have been processed into new cultivars. Regeneration of plants from microspore/pollen provides the most reliable and rapid method to produce haploids, which are extremely valuable in plant breeding and genetics. With haploids, homozygosity can be achieved in a single step, cutting down the breeding period to almost half. This is particularly important for highly heterozygous, long-generation tree species, such as neem. Studies on *in vitro* production of haploids of tropical woody perennials have met with very little success. Pollen plants also provide a unique opportunity to screen gametic variation at sporophytic level. This approach has enabled selection of several gametoclones, which could be developed into new cultivars. Even the triploid cells of endosperm are totipotent, which provides a direct and easy approach to regenerate triploid plants difficult to raise *in vivo*.

A rose grown from tissue culture.

Plant tissue culture is a collection of techniques used to maintain or grow plant cells, tissues or organs under sterile conditions on a nutrient culture medium of known composition. Plant tissue culture is widely used to produce clones of a plant in a method known as micropropagation. Different techniques in plant tissue culture may offer certain advantages over traditional methods of propagation, including:

- The production of exact copies of plants that produce particularly good flowers, fruits, or have other desirable traits.

- To quickly produce mature plants.

- The production of multiples of plants in the absence of seeds or necessary pollinators to produce seeds.

- The regeneration of whole plants from plant cells that have been genetically modified.

- The production of plants in sterile containers that allows them to be moved with greatly reduced chances of transmitting diseases, pests, and pathogens.

- The production of plants from seeds that otherwise have very low chances of germinating and growing, i.e. orchids and *Nepenthes*.

- To clear particular plants of viral and other infections and to quickly multiply these plants as 'cleaned stock' for horticulture and agriculture.

Plant tissue culture relies on the fact that many plant cells have the ability to regenerate a whole plant (totipotency). Single cells, plant cells without cell walls (protoplasts), pieces of leaves, stems or roots can often be used to generate a new plant on culture media given the required nutrients and plant hormones.

Techniques

Preparation of plant tissue for tissue culture is performed under aseptic conditions under HEPA filtered air provided by a laminar flow cabinet. Thereafter, the tissue is grown in sterile containers, such as petri dishes or flasks in a growth room with controlled temperature and light intensity. Living plant materials from the environment are naturally contaminated on their surfaces (and sometimes interiors) with microorganisms, so their surfaces are sterilized in chemical solutions (usually alcohol and sodium or calcium hypochlorite) before suitable samples (known as explants) are taken. The sterile explants are then usually placed on the surface of a sterile solid culture medium, but are sometimes placed directly into a sterile liquid medium, particularly when cell suspension cultures are desired. Solid and liquid media are generally composed of inorganic salts plus a few organic nutrients, vitamins and plant hormones. Solid media are prepared from liquid media with the addition of a gelling agent, usually purified agar.

The composition of the medium, particularly the plant hormones and the nitrogen source (nitrate versus ammonium salts or amino acids) have profound effects on the morphology of the tissues that grow from the initial explant. For example, an excess of auxin will often result in a proliferation of roots, while an excess of cytokinin may yield shoots. A balance of both auxin and cytokinin will often produce an unorganised growth of cells, or callus, but the morphology of the outgrowth will depend on the plant species as well as the medium composition. As cultures grow, pieces are typically sliced off and subcultured onto new media to allow for growth or to alter the morphology of the culture. The skill and experience of the tissue culturist are important in judging which pieces to culture and which to discard.

In vitro tissue culture of potato explants

As shoots emerge from a culture, they may be sliced off and rooted with auxin to produce plantlets which, when mature, can be transferred to potting soil for further growth in the greenhouse as normal plants.

Regeneration Ways

Plant tissue cultures being grown at a USDA seed bank, the National Center for Genetic Resources Preservation.

The specific differences in the regeneration potential of different organs and explants have various explanations. The significant factors include differences in the stage of the cells in the cell cycle, the availability of or ability to transport endogenous growth regulators, and the metabolic capabilities of the cells. The most commonly used tissue explants are the meristematic ends of the plants like the stem tip, axillary bud tip and root tip. These tissues have high rates of cell division and either concentrate or produce required growth regulating substances including auxins and cytokinins.

Shoot regeneration efficiency in tissue culture is usually a quantitative trait that often varies between plant species and within a plant species among subspecies, varieties, cultivars, or ecotypes. Therefore, tissue culture regeneration can become complicated especially when many regeneration procedures have to be developed for different genotypes within the same species.

The three common pathways of plant tissue culture regeneration are propagation from preexisting meristems (shoot culture or nodal culture), organogenesis and non-zygotic embryogenesis.

The propagation of shoots or nodal segments is usually performed in four stages for mass production of plantlets through in vitro vegetative multiplication but organogenesis is a common method of micropropagation that involves tissue regeneration of adventitious organs or axillary buds directly or indirectly from the explants. Non-zygotic embryogenesis is a noteworthy developmental pathway that is highly comparable to that of zygotic embryos and it is an important pathway for producing somaclonal variants, developing artificial seeds, and synthesizing metabolites. Due to the single cell origin of non-zygotic embryos, they are preferred in several regeneration systems for micropropagation, ploidy manipulation, gene transfer, and synthetic seed production. Nonetheless, tissue regeneration via organogenesis has also proved to be advantageous for studying regulatory mechanisms of plant development.

Choice of Explant

The tissue obtained from a plant to be cultured is called an explant.

Explants can be taken from many different parts of a plant, including portions of shoots, leaves, stems, flowers, roots, single undifferentiated cells and from many types of mature cells provided are they still contain living cytoplasm and nuclei and are able de-differentiate and resume cell division. This has given rise to the concept of totipotentency of plant cells. However this is not true for all cells or for all plants. In many species explants of various organs vary in their rates of growth and regeneration, while some do not grow at all. The choice of explant material also determines if the plantlets developed via tissue culture are haploid or diploid. Also the risk of microbial contamination is increased with inappropriate explants.

The first method involving the meristems and induction of multiple shoots is the preferred method for the micropropagation industry since the risks of somaclonal variation (genetic variation induced in tissue culture) are minimal when compared to the other two methods. Somatic embryogenesis is a method that has the potential to be several times higher in multiplication rates and is amenable to handling in liquid culture systems like bioreactors.

Some explants, like the root tip, are hard to isolate and are contaminated with soil microflora that become problematic during the tissue culture process. Certain soil microflora can form tight associations with the root systems, or even grow within the root. Soil particles bound to roots are difficult to remove without injury to the roots that then allows microbial attack. These associated microflora will generally overgrow the tissue culture medium before there is significant growth of plant tissue.

Some cultured tissues are slow in their growth. For them there would be two options: (i) Optimizing the culture medium; (ii) Culturing highly responsive tissues or varieties. Necrosis can spoil cultured tissues. Generally, plant varieties differ in susceptibility to tissue culture necrosis. Thus, by culturing highly responsive varieties (or tissues) it can be managed.

Aerial (above soil) explants are also rich in undesirable microflora. However, they are more easily removed from the explant by gentle rinsing, and the remainder usually can be killed by surface sterilization. Most of the surface microflora do not form tight associations with the plant tissue. Such associations can usually be found by visual inspection as a mosaic, de-colorization or localized necrosis on the surface of the explant.

An alternative for obtaining uncontaminated explants is to take explants from seedlings which are aseptically grown from surface-sterilized seeds. The hard surface of the seed is less permeable to penetration of harsh surface sterilizing agents, such as hypochlorite, so the acceptable conditions of sterilization used for seeds can be much more stringent than for vegetative tissues.

Tissue cultured plants are clones. If the original mother plant used to produce the first explants is susceptible to a pathogen or environmental condition, the entire crop would be susceptible to the same problem. Conversely, any positive traits would remain within the line also.

Applications

Plant tissue culture is used widely in the plant sciences, forestry, and in horticulture. Applications include:

- The commercial production of plants used as potting, landscape, and florist subjects, which uses meristem and shoot culture to produce large numbers of identical individuals.

- To conserve rare or endangered plant species.

- A plant breeder may use tissue culture to screen cells rather than plants for advantageous characters, e.g. herbicide resistance/tolerance.

- Large-scale growth of plant cells in liquid culture in bioreactors for production of valuable compounds, like plant-derived secondary metabolites and recombinant proteins used as biopharmaceuticals.

- To cross distantly related species by protoplast fusion and regeneration of the novel hybrid.

- To rapidly study the molecular basis for physiological, biochemical, and reproductive mechanisms in plants, for example in vitro selection for stress tolerant plants.

- To cross-pollinate distantly related species and then tissue culture the resulting embryo which would otherwise normally die (Embryo Rescue).

- For chromosome doubling and induction of polyploidy, for example doubled haploids, tetraploids, and other forms of polyploids. This is usually achieved by application of antimitotic agents such as colchicine or oryzalin.

- As a tissue for transformation, followed by either short-term testing of genetic constructs or regeneration of transgenic plants.

- Certain techniques such as meristem tip culture can be used to produce clean plant material from virused stock, such as potatoes and many species of soft fruit.

- Production of identical sterile hybrid species can be obtained.

Laboratories

Although some growers and nurseries have their own labs for propagating plants by the technique of tissue culture, a number of independent laboratories provide custom propagation services. The Plant Tissue Culture Information Exchange lists many commercial tissue culture labs. Since plant tissue culture is a very labour-intensive process, this would be an important factor in determining which plants would be commercially viable to propagate in a laboratory.

General Terminology

Plasticity and totipotency: Plants due to their long-life cycle and lack of mobility have developed greater capacity to survive extreme conditions. Many of the processes involved in plant growth and development adapt to environmental conditions. This plasticity allows plants to alter their metabolism, growth and development to best suit their environment. Plants also have, in many cases, the ability to regenerate lost organs (roots, shoots etc.). Cell division can be induced from almost all plant tissues. When plant cells and tissues are cultured in vitro, they generally exhibit a very high degree of plasticity, which allows one type of tissues or organs to be initiated from another type under the influence of chemical stimuli. In this way, whole plants can be subsequently regenerated. For example, embryos may be developed in vitro from somatic cells and haploid cells, as well as from normal zygote and all these, in turn, could develop into whole plants.

Totipotency is the potentiality or property of a cell to produce a whole organism or whole parent plant in the presence of correct physical and chemical stimulus.

Recalcitrant: The opposite of totipotency is recalcitrant. An explant is said to be recalcitrant if it is difficult to give rise to organism or plant.

Explant: A plant organ or piece of tissue used to initiate a culture.

Culture: Growing cells, tissues, plant organs, or whole plants in nutrient medium, under aseptic conditions e.g. cell culture, embryo culture, shoot-tip culture, anther culture.

Contaminants: In tissue culture it refers to the micro-organisms (Bacteria, Fungi), which may inhibit the growth of cells or tissues in culture.

Node: A region on the stem from which a leaf bearing an axillary bud arises.

Morphogenesis: The anatomical and physiological events involved in the growth and development of an organism resulting in the formation of its characteristic organ and structures, or in regeneration.

Meristem: A localized group of actively dividing cells, from which permanent tissue system i.e. root, shoot, leaf and flower are derived. Apical meristem is located at the apices of main and lateral shoots.

Meristematic: having the characteristics of a meristem, especially high mitotic activity.

Meristemoid: A localized group of meristematic cells that arise in the callus and may give rise to roots and or shoots.

Regeneration: In tissue culture, a morphogenetic response that results in the formation of new organs, embryos or whole plants from cultured explants.

Dedifferentiation: The phenomenon of mature cells reverting to meristematic state to produce callus is dedifferentiation. Dedifferentition is possible because the non-dividing quiescent cells of the explant, when grown in a suitable culture medium revert to meristematic state.

Redifferentiation: The ability of the callus cells to differentiate into a plant organ or a whole plant is regarded as redifferentiation.

Adventitious: Developing from unusual points of origin, such as shoots or roots arising from a leaf or stem tissues other than the axils or apex; and embryos from any cell other than the zygote.

Determined: Cells that are committed to a particular pathway of development or differentiation but which have yet to overtly express this pathway.

Heterotrophic: Dependent on an external energy source; not self-reliant compare to autotrophic.

Ploidy: Term used to describe the number of genomes present in the nucleus of a cell or plant.

Polarity: In plants, like in animals, the axes appear very early in development and mostly they are polar in nature. The gradation or change in character occurs along the axis from one end to the other and the condition is referred as 'polarity'. It is visible as morphological differentiation during the development of shoots and roots or is invisible, physiological effect which is expressed during reactivity of cells, tissues and organs in determining cell division and cell growth, and to geotropic or phototropic stimuli. The entire plant is bipolar in nature consisting of two ends, 'plumular' end (where the shoots develop) and 'radicular' (where the roots develop). Besides, there are two other terms, 'distal' and 'proximal'. Distal refers to the part of the plant which is furthest from the original point of attachment i.e. the tip of the leaf or shoot or root, while proximal means nearest to the point of attachment.

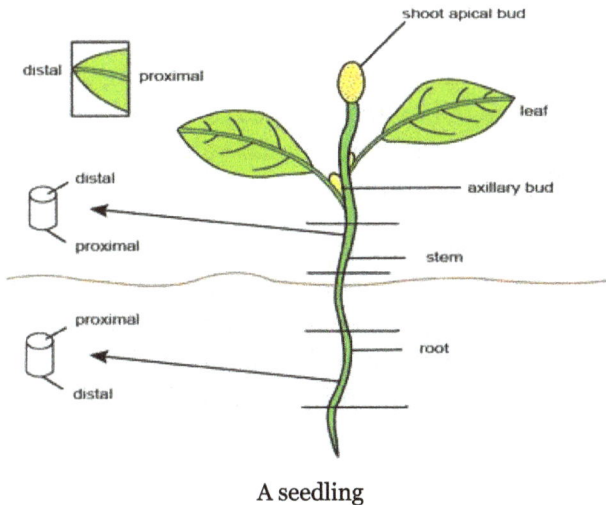

A seedling

Applications of Plant Tissue Culture

- Fundamental biological studies: These can provide a simple, easily manipulated system for investigating a range of phenomena, e.g. the use of cell suspension cultures for studies of cell division.

- High value biochemicals: Many of the natural plant metabolites are required by the pharmaceutical and cosmetic industries. Since environmental conditions can be tightly controlled in vitro, a constant output might be maintained without seasonal variation. Moreover, simplicity of cell culture may allow high yields to be produced at low cost.

- Plant multiplication by micropropagation: Micropropagation offers a rapid means of vegetative (asexual) multiplication. It is important in those plants which are otherwise difficult to propagate, have a high value or where speed of propagation is important. It can also help to raise disease-free stock of a crop by apical meristem culture where the entire plant is infected with pathogenic bacteria or viruses.

- Embryo rescue: Hybrid embryos, formed by the fusion of gametes from distant relatives, frequently abort naturally because the endosperm is not compatible with embryo. Under in vitro conditions, in the presence of correct nutritional medium, these embryos can be developed and form a plant.

- Pollen Culture: The pollen of many species can be induced to develop into entire plant without participating into fertilization process. These plants contain single set of chromosome, as that of gametes, and are haploid in nature. The chromosome set of these haploids can be diploidized by mutagenic treatment like, colchicine, to develop homozygous diploid lines or pure breeding lines. This technique is very useful in case of highly heterozygous tree species with long generation cycle.

- Production of plants with novel characteristics: The somaclonal or gametoclonal variations induced in *in vitro* cultures can be utilized to raise new improved varieties which may have of commercial value such as, new flower color, big canopy plants, large sized grains etc.

- Production of transgenic plants: Characteristics like, insect resistance, that could not be achieved by conventional plant breeding methods, can be transferred from one plant species to another by *Agrobacterium*- mediated gene transfer or by gene transfer methods like, particle gun method (biolistics), electroporation, microinjection and polyethylene glycol-mediated gene transfer.

Protoplast fusion: The protoplasts can be fused to form somatic hybrids. Such fusion products are the result of the union of two or more protoplasts from similar or dissimilar parents.

Laboratory Design and Development

The size of tissue culture lab and the amount and type of equipment used depend upon the nature of the work to be undertaken and the funds available. A standard tissue culture laboratory should provide facilities for:

- washing and storage of glassware, plasticware

- preparation, sterilization and storage of nutrient media

- aseptic manipulation of plant material

- maintenance of cultures under controlled temperature, light and humidity

- observation of cultures, data collection and photographic facility

- acclimatization of in vitro developed plants. The overall design must focus on maintaining aseptic conditions.

At least three separate rooms should be available one for washing up, storage and media preparation (the media preparation room); a second room, containing laminar-air-flow or clean air cabinets for dissection of plant tissues and subculturing (dissection room or sterilization room); and the third room to incubate cultures (culture room). This culture room should contain a culture observation table provided with binoculars or stereozoom microscope and an adequate light source. Additionally, a green house facility is required for hardening-off in vitro plantlets. For a commercial set-up, a more elaborate set-up is required. A suitable floor plan is shown in figure.

Media Preparation Room

The washing area in the media room should be provided with brushes of various sizes and shapes, a large sink, preferably lead-lined to resist acids and alkalis, and running hot and cold water. It should also have large plastic buckets to soak the labware to be washed in detergent, hot-air oven to dry washed labware and a dust-proof cupboard to store them. If the preparation of the medium and washing of the labware are done in the same room, a temporary partition can be constructed between the two areas to guard any interference in the two activities. A continuous supply of water is essential for media preparation and washing of labware. A water distillation unit of around 2 litre/h, a Milli-Q water purification systems needs to be installed.

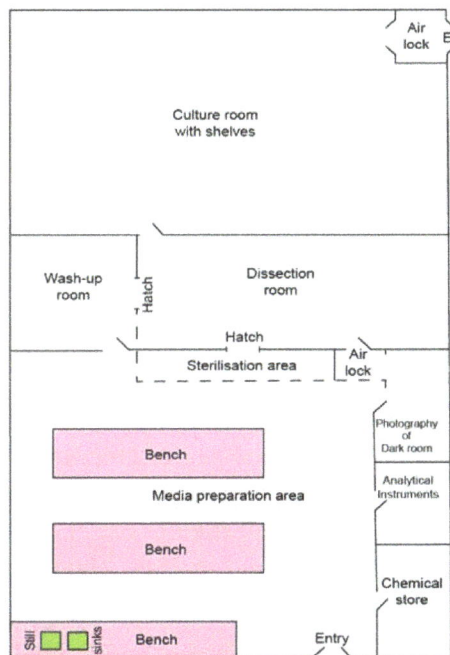

A floor plan for plant tissue culture laboratory

Culture Room

The room for maintaining cultures should be maintained at temperature 25 ±2°C, controlled by air conditioners and heaters attached to a temperature controller are used. For higher or lower temperature treatments, special incubators with built-in fluorescent light can be used outside the culture room. Cultures are generally grown in diffuse light from cool, white, fluorescent tubes. Lights can be controlled with automatic time clocks. Generally, a 16-hour day and 8-hour nights are used. The culture room requires specially designed shelving to store cultures. Some laboratories have shelves along the walls, others have them fitted onto angle-iron frames placed in a convenient position. Shelves can be made of rigid wire mesh, wood or any building material that can be kept clean and dust-free. Insulation between the shelf lights and the shelf above will ensure an even temperature around the cultures. While flasks, jars and petridishes can be placed directly on the shelf or trays of suitable sizes, culture tubes require some sort of support. Metallic wire racks or polypropylene racks, each with a holding capacity of 18-24 tubes, are suitable for the purpose.

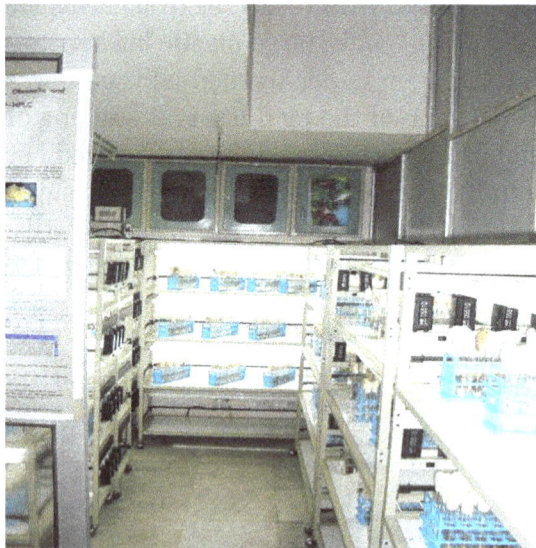

Culture room facility with humidifier, timer, wall cabinets, illuminated trolleys
and test-tube racks

Dissection Room or Sterilization Room

This area should have restricted entry, which is needed to ensure the sterile conditions required for the transfer operations. For sterile transfer operations, the laminar-air-flow cabinets are used. Temperature control is essential in this room as the heat is produced continuously from the flames of burners in the hoods. The room should be constructed in a way to minimize the dust particles and for easy cleaning. Several precautions can be taken including the removal of shoes before entering the area.

The laminar horizontal flow sterile transfer cabinets are available in various sizes from many commercial sources. They should be designed with horizontal air flow from the back to the front, and equipped with gas cocks if gas burners are to be used. Electrical outlets are needed for use of electric sterilizers and microscopes, and if weighing is to be done in the hoods. A stainless steel working platform is most durable, easy to keep clean and to prevent the unwanted damage due to accidental fire. Sometimes it is fitted with Ultraviolet light to maintain sterility inside the cabinet. UV light is a source of ozone, which can be mutagenic, therefore, utmost care is to be taken while using this. Although UV light is not necessary, a short exposure time of 3-5 min to cabinet is fine sometimes. Work can be started after 10-15 min of switching on the air flow, and one can work uninterrupted for long hours.

A Laminar-air-flow cabinet has small motor to blow air which first passes through a coarse filter, where it loses large particles, and subsequently through a fine filter known as 'high efficiency particulate air (HEPA). The HEPA filters remove particles larger than 0.3 μm, and the ultraclean air flows through the working area. The velocity of the ultra clean air is about 27 ± 3 m min^{-1} which is adequate for preventing the contamination of the working area as long as the flow is on. The flow of the air does not in any way hamper the use of a spirit lamp or a Bunsen burner. A

Laminar-air-hood with coarse filter, HEPA filter, gas cock, gas cylinder and electrical outlets

Greenhouse

The greenhouse facility is required to grow parent pants and to acclimatize in vitro raised plantlets. The size and facility inside the green house vary with the requirement and depends on the funds available with the laboratory. However, minimum facilities for maintaining humidity by fogging, misting or a fan and pad system, reduced light, cooling system for summers and heating system for winters must be provided. It would be desirable to have a potting room adjacent to this facility.

Equipments and Apparatus

Media preparation area:

- benches at a height suitable to work while standing

- pH meter is used to determine the pH of various media used for tissue culture. pH indicator paper can also be used for the purpose but it is less accurate. The standard media pH is maintained at 5.8.

- hot-plate-cum-magnetic stirrer for dissolving chemicals and during media preparation

- an autoclave or domestic pressure cooker is crucial instrument for a tissue culture laboratory. High pressure heat is needed to sterilize media, water, labware, forceps, needles etc. Certain spores from fungi and bacteria can only be killed at a temperature of 121°C and 15 pounds per square inch (psi) for 15-20 min. A caution should be taken while opening the door of autoclave and it should be open when the pressure drops to zero. Opening the door immediately can lead to a rapid change in the temperature, resulting in breakage of glassware and steam burning of operator.

- plastic carboys for storing distilled water required for media preparation and final washing of labware.

- balances near dry corner of the media room. High quality microbalance are required to weigh smallest of the quantities. Additionally a top pan balance is required for less sensitive quantities.

- hot-air oven to keep autoclaved medium warm before pouring into vessels. It is also used for the dry heat sterilization of clean glassware like, Petridishes, culture tubes, pipettes etc. Typical sterilizing conditions are 160-170 °C/1hr.

- Dishwasher for cleaning glass pipettes in running water

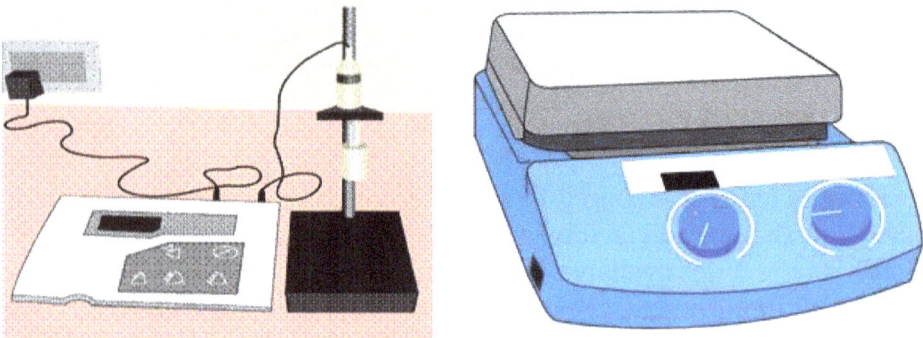

pH Meter and Magnetic stirrer-cum-heater

Autoclave with accessories and High Precision Weighing Balance

Storage Area

- a deep freezer (-20°C to -80°C) / refrigerator for storage of enzyme solutions, stock solutions plant materials and all temperature-sensitive chemicals.

- microwave oven to melt agar solidified media

- Upright and inverted light microscope with camera attachment for recording the morphogenic responses from various explants, calli, cells and protoplasts. Inverted microscope gives the clear views of cultures settled at the bottom of Petridishes.

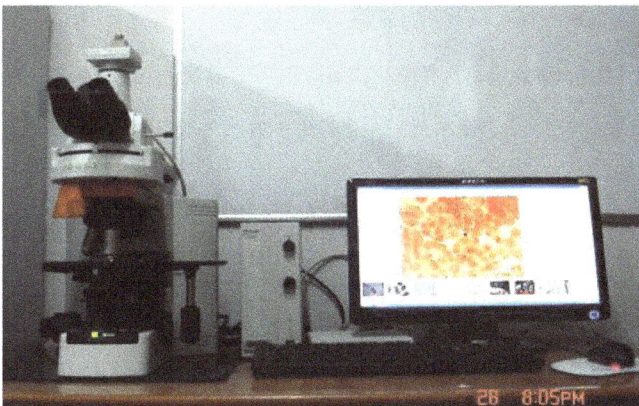
Upright light microscope with CCD camera attachment

Dissection Room

- laminar-air-flow cabinet within which tissue culture work can be carried out under sterilized environment

- glass bead sterilizer where temperature of beads is raised to 250°C in 15-20 min with 15 s cut off. Here the sterilization of instruments is effecting by pushing them into the beads for 5-7 s. This is much safer compare to the Bunsen burner heating of instruments like, forceps, needles, scalpels etc.

- binocular microscope to observe surface details and morphogenic responses of cultures and their possible contamination.

- low speed table-top centrifuge to sediment cells or protoplasts

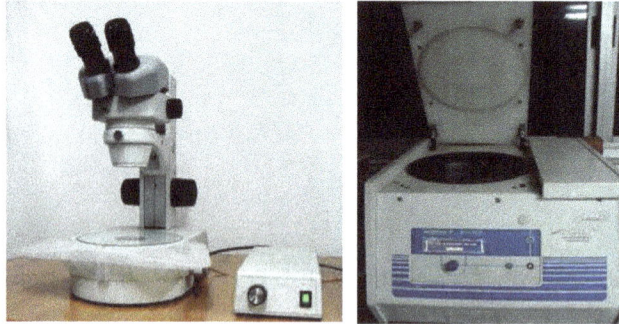

Binocular stereozoom microscope (left side) and Centrifuge (right figure)

Culture Room

- air (or heating / cooling system) to maintain 25±2 °C temperature

- racks for holding test-tubes

- lights to provide diffuse light and to maintain photoperiod

- shakers with various sized clamps for different sized flasks to grow cells in liquid medium

- thermostat and time clock for lights

- wall cabinets for dark incubation of cultures

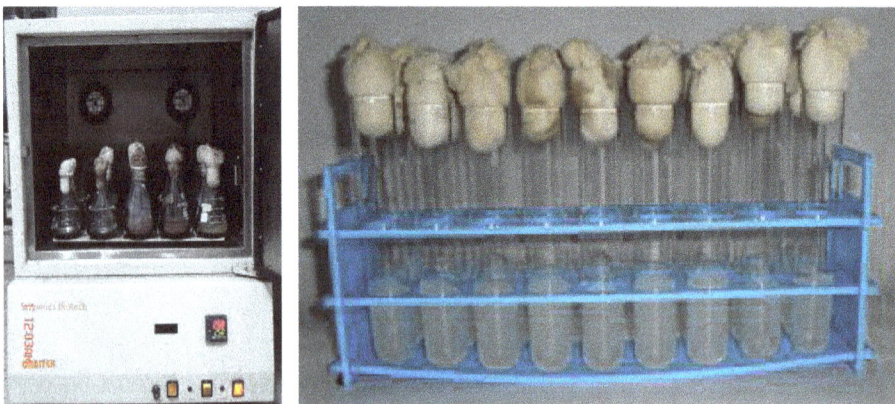

Incubator shaker and Test-tube racks

Other Apparatus

- beakers (100 mL, 250 mL, 1 L, 5 L)

- measuring cylinders (5 mL, 10 mL, 25 mL, 50 mL, 100 mL, 500 mL, 1L, 2 L, 5 L)

- graduated pipettes and teats

- reagent bottles for storing liquid chemicals and stock solutions (glass or plastic)

- culture tubes and flasks (glass or polypropylene or disposable)

- plastic baskets

- filter membrane, preferably nylon, of sizes 0.22 µm and 0.45 µm, holders and hypodermic syringes (for solutions requiring filter sterilization)

- large forceps (blunt and fine points) and scalpels for dissecting and subculturing plant material.

- Scalpel handles (no. 3) and blades (no. 11)

- Chemicals and reagents for preparing culture media

- Disposable gloves and masks.

- Micropipettes of maximum volume size 5000 µL, 1000 µL, 500 µL, 250 µL, 100 µL

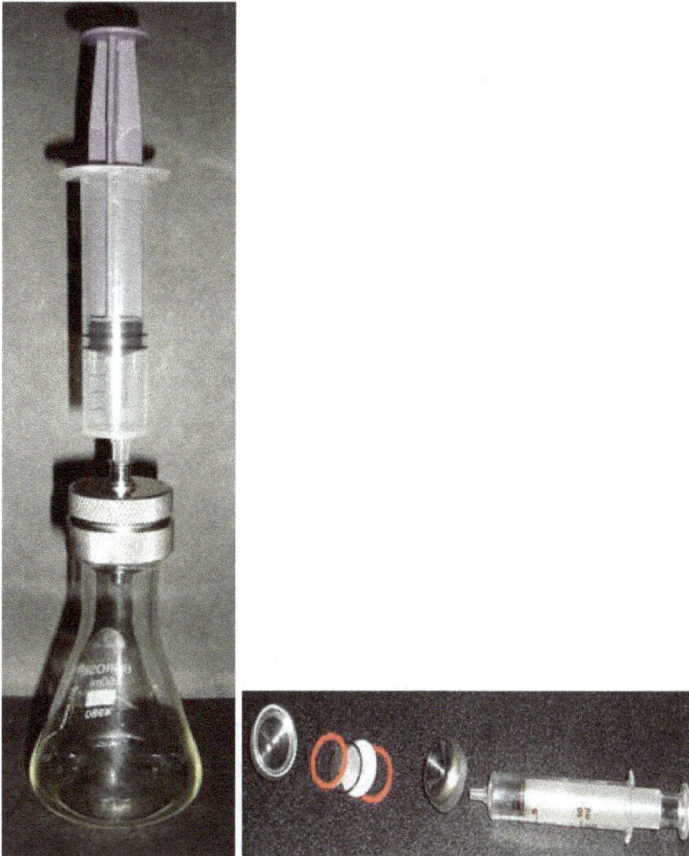

(A) Syringe with filter assembly fitted on conical flask, (B) Disassembled filter assembly

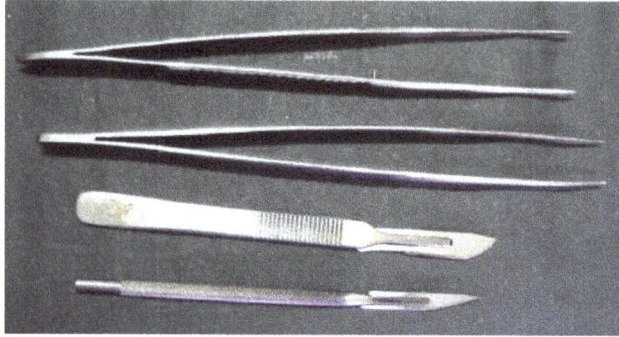
Forceps and scalpels for dissection

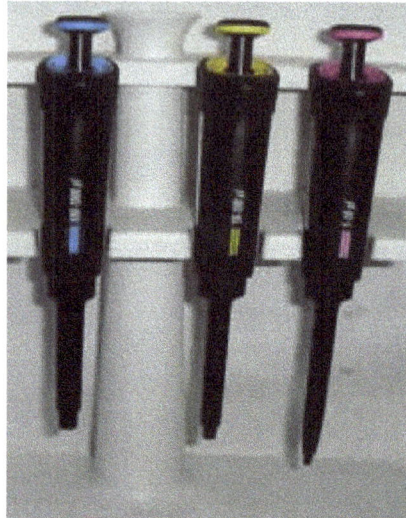
Micropipettes .

Plant Tissue Culture Techniques

Growing any part of the plant (explants) like, cells, tissues and organs, in an artificial medium under controlled conditions (aseptic conditions) for obtaining large scale plant propagation is called micropropagation. The basic concept of micropropagation is the plasticity, totipotency, differentiation, dedifferentiation and redifferentiation, which provide the better understanding of the plant cell culture and regeneration. Plants, due to their long life span, have the ability to withhold the extremes of conditions unlike animals. The plasticity allows plants to alter their metabolism, growth and development to best suit their environment. When plant cells and tissues are cultured *in vitro* , they generally exhibit a very high degree of plasticity, which allows one type of tissue or organ to be initiated from another type. Hence, whole plants can be subsequently regenerated and this regenerated whole plant has the capability to express the total genetic potential of the parent plant. This is unique feature of plant cells and is not seen in animals. Unlike animals, where differentiation is generally irreversible, in plants even highly mature and differentiated cells retain the ability to regress to a meristematic state as long as they have an intact membrane system and a viable nucleus. However, sieve tube elements and xylem elements do not divide any more where the nuclei have started to disintegrate, According to Gautheret (1966) the degree of regression a cell can undergo would depend on the cytological and physiological state of the cell. The meristematic tissues are differentiated into simple or complex tissues called differentiation. Reversion of mature tissues into meristematic state leading to the formation of callus is called dedifferentiation. The ability of callus to develop into shoots or roots or embryoid is called redifferentiation. The inherent potentiality of a plant cell to give rise to entire plant and its capacity is often retained even after the cell has undergone final differentiation in the plant system is described as cellular totipotency.

Micropropagation Vs. Conventional Method of Propagation

All living plant cells, irrespective of their nature of specialization and ploidy level, have been shown to regenerate plants *via* organogenesis or embryogenesis. The latter involves a highly specialized mode of development that normally occurs only inside the seed, under the cover of several layers of parental tissues. Consequently, the observation of developing embryos and their isolation in intact and living conditions for experimental studies have been extremely difficult. *In vitro* production of embryos from somatic and gametic cells has opened up the possibility of obtaining large numbers of embryos of different stages, enabling investigations on cellular, genetic and physiological control of embryogenesis (induction, pattern formation, organ differentiation and maturation). *In vitro* expression of cellular totipotency and other techniques of plant tissue culture have also facilitated and/ or accelerated the traditional methods of plant improvement, propagation and conservation.

Micropropagation Vs. Vegetative Propagation

The vegetative propagation has been conventionally used to raise genetically uniform large scale plants for thousands of years. However, this technique is applicable to only limited number of species. In contrast to this, micropropagation has several advantages which are summarized here:

i. The rapid multiplication of species difficult to multiply by conventional vegetative means. The technique permits the production of elite clones of selected plants.

ii. The technique is independent of seasonal and geographical constraints.

iii. It enable large numbers of plants to be brought to the market place in lesser time which results in faster return on the investment that went into the breeding work.

iv. To generate disease-free (particularly virus-free) parental plant stock.

v. To raise pure breeding lines by *in vitro* haploid and triploid plant development in lesser time.

vi. It can be utilized to raise new varieties and preservation of germplasm

vii. It offers constant production of secondary medicinal metabolites.

Cell Differentiation

During *in-vitro* and *in vivo* cytodifferentiation (cell differentiation), the main emphasis has been on vascular differentiation, especially tracheary elements (TEs). These can be easily observed by staining and can be scored in macerated preparations of the tissues. Tissue differentiation goes on in a fixed manner and is the characteristic of the species and the organs

Factors Affecting Vascular Tissue Differentiation

Vascular differentiation is majorly affected qualitatively and quantitatively by two factors, auxin and sucrose. Cytokinins and gibberellins also play an important role in the process of xylogenesis. Depending upon the characteristics of different species, concentration of phytohormones, sucrose and other salt level varies and accordingly it leads to the vascular tissue differentiation.

Micropropagation Techniques

Strategies for Propagation in Vitro

Typical micropropagation system can be broadly divided into five distinct stages:

*   The stage zero is the selection of mother plant and preparation of explant.

The first stage is the initiation of a sterile culture of the explant in a particular enriched medium for specific species.

The second stage includes initiation of cell division from almost any part of the plant system to initiate regeneration or multiplication of shoots or other propagules from the explant. Adventitious shoot proliferation is the most frequently used multiplication technique in micropropagation systems. The culture media and growth conditions used in second stage need to be optimized for maximum rate of multiplication.

The third stage is the development of roots on the shoots to produce plantlets. Specialized media may or may not be required to induce roots, depending upon the species.

The final or the fourth stage is to produce self-sufficient plants. This stage usually involves a hardening-off process and acclimatization of plants in soil under green-house conditions for later transplanting to the field.

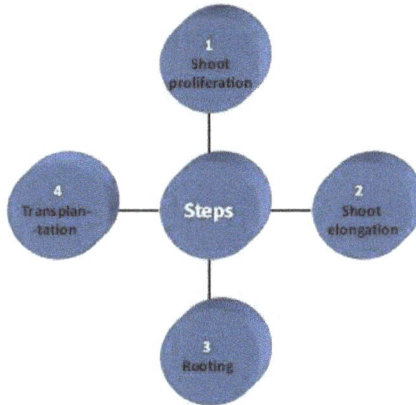

Micropropagation stages

Mode of Differentiation

Regenerants may differentiate either directly from the explants or indirectly via cal-lusing. Dedifferentiation favours unorganized cell growth and the resultant developed callus has meristems randomly distributed in the callus. Most of these meristems, if provided appropriate *invitro* conditions, would differentiate shoot-buds, roots or embryos.

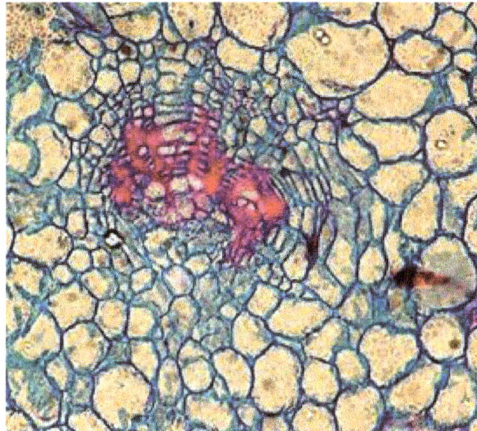

Meristem formation in callus. Thick tracheidal cells (stained red) are surrounded by cambium like cells (stained blue)

Trouble Shooting

- Few explants exude dark colored compounds, like phenols, pigments etc which leach into the medium from the cut ends of the explant. It results in the browning of tissues and the medium as well. The browning of medium is associated with poor culture establishment and low regeneration capacity of the explants. This can be overcome by:

i   minimizing the wounding of explants during isolation and surface disinfection to reduce this browning response.

ii   washing or incubation of explants for 3-5 hrs in sterile distilled water to remove phenolics responsible for browning of medium or explants.

iii frequent subculture of explants with excision to fresh medium at regular intervals.

iv   initial establishment of cultures in liquid medium and later transfer to the semi-solid medium.

v   culture of explants on porous substrate or paper bridges.

vi   addition of activated charcoal (AC) or polyvinylpyrrolidone (PVP) for adsorbtion of phenolics.

vii antioxidants like ascorbic acid, citric acid etc. can also be used to prevent browning of tissues in culture.

- Appearance of vitrified tissues (hyperhydricity), a physiological disorder occurring in the *in vitro* cultures due to which the tissues look transparent and fluffy resulting from excessive intake of water. Hyperhydricity can be caused by a high concentration of cytokinin or low concentration of gelling agent or high water retention capacity of explants if the container is tightly closed.

- Loss of regeneration ability in long-term cultures due to epigenetic variations (temporary variations) and culture aging, including transition from juvenile to mature stage. Epigenetic variation are phenotypic temporary variations which disappear as soon as the culture conditions are removed.

- Genotypic variations are also seen in the cultures, therefore, cytological, biochemical and molecular analyses are required to confirm clonal fidelity of *in vitro* regenerants. Besides, morphological and physiological testing is also required to remove undesired genetic variability.

Tissue Culture Media

Preparation and Handling

The simplest method of preparing media is to use commercially available, dry, powdered media containing mineral elements and growth regulators. By following the procedure written on the packets, dissolve the powder in distilled or demineralized water (10% less than the final volume of the medium). After adding sugar and other desired supplements like, plant growth regulators, make up the final volume with distilled water, adjust the pH, add agar and then autoclave the medium.

An alternative method of media preparation is to prepare a series of concentrated stock solutions which can be combined later as required. For preparing stock solutions and media, use glass-distilled or demineralized water and chemicals of high purity, analytical reagent (AR) grade.

Composition of Widely used Tissue Culture Media

Both the media listed in the below tables can be prepared from stock solutions of:

i. Macronutrients: As its name suggests, in plant tissue culture media these components provide the elements which are required in large amounts (concentrations greater than 0.5 mmole l^{-1}) by cultured plant cells. Macronutrients are usually considered to be carbon, nitrogen, phosphorous, magnesium, potassium, calcium and sulphur.

ii. Micronutrients: It provides the elements that are required in trace amounts (concentrations less than 0.5 mmole l^{-1}) for plant growth and development. These include, manganese, copper, cobalt, boron, iron, molybdenum, zinc and iodine.

iii. Iron source: It is considered the most important constituent and required for the formation of several chlorophyll precursors and is a component of ferredoxins (proteins containing iron) which are important oxidation : reduction reagents.

iv. Organic supplements (vitamins): Like animals, in plants too vitamins provide nutrition for healthy growth and development. Although plants synthesize many vitamins under natural conditions and, therefore, under in vitro conditions they are supplied from outside to maintain biosynthetic capacity of plant cells in vitro. There are no firm rules as to what vitamins are essential for plant tissues and cell cultures. The only two vitamins that are considered to be essential are myo-inositol and thiamine. Myo-inositol is considered to be vitamin B and has many diverse roles in cellular metabolism and physiology. It is also involved in the biosynthesis of vitamin C.

v. Carbon source: This is supplied in the form of carbohydrate. Plant cells and tissues in the culture medium are heterotrophic and are dependent on external source of carbon. Sucrose is the preferred carbon source as it is economical, readily available, relatively stable to autoclaving and readily assimilated by plant cells. During sterilization (by autoclaving) of medium, sucrose gets hydrolyzed to glucose and fructose. Plant cells in culture first utilize glucose and then fructose. Besides sucrose, other carbohydrates such as, lactose, maltose, galactose are also used in culture media but with a very limited success.

Dilutions:

$$\frac{\text{required concentration} \times \text{medium volume}}{\text{concentration of stock solution}} = \text{volume of stock required}$$

Table: The media elements and their functions

S.No.	Elements	Functions
1	Oxygen	Common cell component, electron acceptor.
2	Carbon	Common cellular component, forms basic backbone of most biochemicals.
3	Nitrogen	Part of proteins, vitamins, amino acids and coenzymes.
4	Sulphur	Part of some amino acids and some coenzymes.
5	Potassium	Principal inorganic cation
6	Magnesium	Important coenzyme factor and part of chlorophyll molecules
7	Manganese	Important cofactor
8	Calcium	Important constituent of cell wall and enzyme cofactor
9	Iron	Part of cytochromes
10	Cobalt	Part of some vitamins
11	Copper	Enzyme cofactor
12	Zinc	Enzyme cofactor
13	Molybdenum	Enzyme cofactor

Table: The composition of Gamborg's B_5 Medium (1968)

Component	Concentration in stock ($mg\ l^{-1}$)	Concentration in medium ($mg\ l^{-1}$)	Volume of stock per litre of medium (ml)
Macronutrients			
KNO_3	50000	2500	
$CaCl_2\ 2H_2O$	3000	150	
$(NH_4)_2SO_4$	2680	134	50
$MgSO_4\ 7H_2O$	5000	250	
$NaH_2PO_4\ H_2O$	3000	150	
Micronutrients			
KI	30	0.75	
H_3BO_3	120	3	
$MnSO_4\ 4H_2O$	400	10	
$ZnSO_4\ 7H_2O$	80	2	25
$Na_2MoO_4.2H_2O$	10	0.25	
$CuSO_4\ 5H_2O$	1	0.025	
$CoCl_2\ 6H_2O$	1	0.025	
Iron Source			
FeNaEDTA	3670	36.7	10
Vitamins			
Myo-inositol	Add freshly to the medium	100	
Pyridoxine-HCl	1000	1	
Thiamine-HCl	10000	10	1
Nicotinic acid	1000	1	
Carbon Source			
Sucrose	Add freshly to the medium	$30g\ l^{-1}$	
Adjust pH to 5.5 before autoclaving			

- Stock concentration of macronutrients is for 20 litres of medium, while micronutrients stock is for 40 litres of medium, iron for 100 litres of medium and vitamins stock is for 1000 litres of medium.

- Myoinositol and sucrose are added freshly to the medium.

Table: The composition of Murashige and Skoog (MS) Medium (1962)

Component	Concentration in stock (mg l⁻¹)	Concentration in medium (mg l⁻¹)	Volume of stock per litre of medium (ml)
Macronutrients			
NH_4NO_3	33000	1650	
KNO_3	38000	1900	
$CaCl_2.2H_2O$	8800	440	50
$MgSO_4.7H_2O$	7400	370	
KH_2PO_4	3400	170	
Micronutrients			
KI	166	0.83	
H_3BO_3	1240	6.2	
$MnSO_4.4H_2O$	4460	22.3	
$ZnSO_4.7H_2O$	1720	8.6	5
$Na_2MoO_4.2H_2O$	50	0.25	
$CuSO_4.5H_2O$	5	0.025	
$CoCl_2.6H_2O$	5	0.025	
Iron Source			
$FeSO_4.7H_2O$	5560	27.8	5
$Na_2EDTA.2H_2O$	7460	37.3	
Vitamins			
Myo-inositol	Add freshly to the medium	100	
Nicotinic acid	100	0.5	
Pyridoxine-HCl	100	0.5	5
Thiamine-HCl	100	0.5	
Glycine	400	2	
Carbon Source			
Sucrose	Add freshly to the medium	30g l⁻¹	
Adjust pH to 5.7-5.8 before autoclaving			

- Stock concentration of macronutrients is for 20 litres of medium, while micro-nutrients, iron and vitamins stock concentrations are prepared for 200 litres of medium.

- Myoinositol and sucrose are added freshly to the medium.

- Dissolve 5.56 g of $FeSO_4$.$7H_2O$ in 350 ml of water. Apply heat if needed. Dissolve 7.46 g of Na_2 EDTA in 350 ml of water. Apply heat if needed. When both solutions are dissolved, combine and bring to 1 litre final volume. The chelation reaction is forced to completion by autoclaving. The final stock solution should be deep golden yellow in color.

The steps involved in preparing a medium are summarized below:

√ Add appropriate quantities of various stock solutions, including growth regulators and other special supplements. Make up the final volume of the medium with distilled water.

√ Add and dissolve sucrose.

√ After mixing well, adjust the pH of the medium in the range of 5.5-5.8, using 0.1 N NaOH or 0.1 N HCl (above 6.0 pH gives a fairly hard medium and pH below 5.0 does not allow satisfactory gelling of the agar).

√ Add agar, stir and heat to dissolve. Alternatively, heat in the autoclave at low pressure, or in a microwave oven.

√ Once the agar is dissolved, pour the medium into culture vessels, cap and autoclave at 121°C for 15 to 20 min at 15 pounds per square inch (psi). If using pre-sterilized, non-autoclavable plastic culture vessels, the medium may be autoclaved in flasks or media bottles. After autoclaving, allow the medium to cool to around 60°C before pouring under aseptic conditions.

√ Allow the medium to cool to room temperature. Store in dust-free areas or refrigerate at 7°C (temperature lower than 7°C alter the gel structure of the agar).

Gelling Agents

The media listed above are only for liquids, often in plant cell culture a 'semi-solid' medium is used. To make a semi-solid medium, a gelling agent is added to the liquid medium before autoclaving. Gelling agents are usually polymers that set on cooling after autoclaving.

i.   Agar: Agar is obtained from red algae- *Gelidium amansii* . It is a mixture of polysaccharides. It is used as a gelling agent due to the reasons: (a) It does not react with the media constituents (b) It is not digested by plant enzymes and is stable at culture temperature.

ii. Agarose: It is obtained by purifying agar to remove the agaropectins. This is required where high gel strength is needed, such as in single cell or protoplast cultures.

iii. Gelrite: It is produced by bacterium *Pseudomonas elodea* . It can be readily prepared in cold solution at room temperature. It sets as a clear gel which assists easy observation of cultures and their possible contamination. Unlike agar, the gel strength of gelrite is unaffected over a wide range of pH. However, few plants show hyperhydricity on gelrite due to freely available water.

iv. Gelatin: It is used at a high concentration (10%) with a limited success. This is mainly because gelatin melts at low temperature (25°C) and as a result the gelling property is lost.

Plant Growth Regulators

In addition to nutrients, four broad classes of growth regulators, such as, auxins, cytokinins, gibberellins and abscisic acid are important in tissue culture. In contrast with animal hormones, the synthesis of a plant growth regulator is often not localized in a specific tissue but may occur in many different tissues. They may be transported and act in distant tissues and often have their action at the site of synthesis. Another property of plant growth regulators is their lack of specificity- each of them influences a wide range of processes.

The growth, differentiation, organogenesis and embryogenesis of tissues become feasible only on the addition of one or more of these classes of growth regulators to a medium. In tissue culture, two classes of plant growth regulators, cytokinins and auxins, are of major importance. Others, in particular, gibberellins, ethylene and abscisic acid have been used occasionally. Auxins are found to influence cell elongation, cell division, induction of primary vascular tissue, adventitious root formation, callus formation and fruit growth. The cytokinins promote cell division and axillary shoot proliferation while auxins inhibit the outgrowth of axillary buds. The auxin favours DNA duplication and cytokinins enable the separation of chromosome. Besides, cytokinin in tissue culture media, promote adventitious shoot formation in callus cultures or directly from the explants and, occasionally, inhibition of excessive root formation and are, therefore, left out from rooting media. The ratio of plant growth regulators required for root or shoot induction varies considerably with the tissue and is directly related to the amount of growth regulators present at endogenous levels within the explants. In general, shoots are formed at high cytokinin and low auxin concentrations in the medium, roots at low cytokinin and high auxin concentrations and callus at intermediate concentrations of both plant growth regulators. Commonly used plant growth regulators are listed in table.

Table: Stock solutions of growth regulators

Compound	Abbreviations	mg/50 ml (1 mM or 10^{-3} Molar)
CYTOKININS		
6-Benzyladenine	BA	11.25
N^6-(2-isopentenyl) adenine	2-iP	10.15
6-Furfurylaminopurine	Kinetin	10.75
Zeatin	ZEA	10.95
Thidiazuron	TDZ	11.00
Note: Dissolve cytokinins in few drops of 1 N NaOH; stir; heat gently and make to volume. TDZ is dissolved in 95% ethanol.		
AUXINS		
Indole-3-acetic acid	IAA	8.76
Indole-3-butyric acid	IBA	10.16
α-Naphthaleneacetic acid	NAA	9.31
2,4-Dichlorophenoxyacetic acid	2,4-D	11.05
2,4,5-Trichlorophenoxyacetic acid	2,4,5-T	12.78
p-Chlorophenoxyacetic acid	4-CPA	9.33
Picloram	PIC	12.06
Note: Dissolve auxins in 95% ethanol or 1N NaOH, stir, heat gently; gradually add water to volume. Dissolve picloram in DMSO.		
OTHERS		
Silver Nitrate	$AgNO_3$	9.00
Gibberellic acid	GA_3	17.32
Abscisic acid	ABA	13.20
Note: Dissolve in 95% ethanol or 1N NaOH, stir, heat gently; gradually add water to volume.		

1 molar = the molecular weight in g/l

1 mM = the molecular weight in mg/l

ppm = parts per million = mg/l

Establishing Aseptic Cultures

Plant tissue culture media contain sugar and so support the growth of many micro-organisms (bacteria and fungi). When these microorganisms reach a medium, they generally grow much faster than the cultured plant materials. Their growth and toxic metabolites will affect, and may even kill, the tissue cultures. It is, therefore, essential to maintain a completely aseptic environment inside the culture vessels.

There are several possible sources of contamination of the medium:

- the culture vessel
- the medium itself
- the explant (plant tissue)
- the environment of the transfer area
- the instruments used to handle plant material during establishment and subculture
- the environment of the culture room.

Autoclaving media will eliminate contamination from the culture vessel or the medium. In some cases, substances such as gibberellic acid, abscisic acid (ABA), urea and certain vitamins are thermolabile and break down upon autoclaving. These chemicals can be sterilized by membrane filtration using microfilters of pore size 0.22-0.45 μm which is suitable enough to exclude pathogens. Later the filter sterilized compound can be added to autoclaved medium cooled to around 40°C.

To prevent the environment of the culture room from being the source of contamination, keep the culture room as dust- free as possible and remove contaminated cultures from the area as soon as they are detected. Ideally, the culture room should be clean, filtered air which has passed through high efficiency particulate air (HEPA) filters.

The transfer area in most laboratories is within a laminar air-flow cabinet. A laminar air-flow cabinet has a small fan which blows air through a coarse filter to remove large dust particles and then through a fine HEPA filter to remove microbes, their spores and other particles larger than 0.3 μm. The velocity of the air coming out of the fine filter is about 27 ± 3 m/min, which keeps airborne microorganisms out of the working area. The working area is swabbed with 70% alcohol (or equivalent) and instruments dipped in 70% alcohol, flamed and cooled before use.

Caution : Prolonged contact with alcohol can cause skin irritation, and other health problems can result from the inhalation of fumes. Use ethanol rather than methanol, and surgical gloves when handling. Take care with ultraviolet light as it can permanently damage eyes and promote skin cancer. Laminar flow cabinets equipped with ultraviolet light for surface sterilization should be fitted with safety doors which can be closed when ultraviolet light is used.

Plant surfaces carry a wide range of microorganisms. The tissue must be thoroughly sur-face-sterilized before being placed on the nutrient medium. Discard cultures with fun-gal or bacterial contamination. Solutions of sodium or calcium hypochlorite are usually effective in disinfecting plant tissues. Placing tissues in a 0.5 to 1% solution of sodium hypochlorite for 10 to 15 minutes will disinfect most tissues. Surface sterilants are toxic to plant tissues. Choose the concentration of the sterilizing agent and the length of time to minimize tissue damage, which shows up as white, bleached areas. Other techniques for surface sterilisation include dipping plant material for a few seconds in 90% ethanol or placing in running water for 30 minutes and 2 hours before disinfection.

Caution : Take care with powdered calcium hypochlorite as it is a powerful reducing agent. If calcium hypochlorite is stored moist and the container opened later, it can explode. Store calcium hypochlorite in a sealed container in a dry place.

A summary of the six steps commonly involved in establishing and maintaining aseptic plant tissue culture follows:

i.   Collect pieces of plant material (ex-plants) in a screw-cap bottle. Immerse them in a dilute solution of the disinfectant containing a wetting agent. Replace the lid and store the bottle in the laminar air flow cabinet. Shake the bottle two or three times during the sterilization period.

ii. Remove the lid and drain carefully. Thoroughly rinse the plant material in sterilized distilled water and replace the lid. After shaking a few minutes, discard the water. Rinse two or three times more.

iii. Transfer the material to a pre-sterilized Petri-dishes or test-tubes.

iv. Sterilize the required instruments by dipping them in 70% ethanol and flamed them. Allow to cool. Sterilize the instruments after each time they are used to handle tissue.

v.   Prepare suitable explants from the surface sterilized material using sterilized instruments (scalpels, needles, forceps, etc.).

vi. Quickly remove the lid of the culture vessel, transfer the explants on to the me-dium, flame the neck of the vessel (only if glass) and replace the lid.

If handling aseptic plant materials during routine subculture, omit the first two steps.

References

- Aina, O; Quesenberry, K.; Gallo, M (2012). "In vitro induction of tetraploids in Arachis paraguar-iensis". Plant Cell, Tissue and Organ Culture (PCTOC). 111: 231–238. doi:10.1007/s11240-012-0191-0

- Sathyanarayana, B.N. (2007). Plant Tissue Culture: Practices and New Experimental Protocols. I. K. International. pp. 106–. ISBN 978-81-89866-11-2

- Mukund R. Shukla; A. Maxwell P. Jones; J. Alan Sullivan; Chunzhao Liu; Susan Gosling; Praveen K. Saxena (April 2012). "In vitro conservation of American elm (Ulmus americana): potential role of auxin metabolism in sustained plant proliferation". Canadian Journal of Forest Research. 42 (4): 686–697. doi:10.1139/x2012-022

- Vasil, I.K.; Vasil, V. (1972). "Totipotency and embryogenesis in plant cell and tissue cultures.". In Vitro. 8: 117–125. doi:10.1007/BF02619487

- Bhojwani, S. S.; Razdan, M. K. (1996). Plant tissue culture: theory and practice (Revised ed.). Elsevier. ISBN 0-444-81623-2

- Pazuki, Arman & Sohani, Mehdi (2013). "Phenotypic evaluation of scutellum-derived calluses in 'Indica' rice cultivars" (PDF). Acta Agriculturae Slovenica. 101 (2): 239–247. doi:10.2478/acas-2013-0020

Micropropagation and Plant Embryogenesis

With many plants not being able to produce seeds or lacking the ability to go through vegetative reproduction, techniques such as micropropagation is used to produce huge amount of plantlets. To heed to the need of producing large-scale seeds, artificial seed production is utilized. This chapter discusses the methods of micropropagation and artificial seed production in a critical manner providing key analysis to the subject matter.

Micropropagation

Micropropagation is the practice of rapidly multiplying stock plant material to produce a large number of progeny plants, using modern plant tissue culture methods.

Micropropagation is used to multiply noble plants such as those that have been genetically modified or bred through conventional plant breeding methods. It is also used to provide a sufficient number of plantlets for planting from a stock plant which does not produce seeds, or does not respond well to vegetative reproduction.

Cornell University botanist Frederick Campion Steward discovered and pioneered micropropagation and plant tissue culture in the late 1950s and early 1960s.

Stages of Micropropagation

Establishment

Micropropagation begins with the selection of plant material to be propagated. The plant tissues are removed from an intact plant in a sterile condition. Clean stock materials that are free of viruses and fungi are important in the production of the healthiest plants. Once the plant material is chosen for culture, the collection of explant(s) begins and is dependent on the type of tissue to be used; including stem tips, anthers, petals, pollen and other plant tissues. The explant material is then surface sterilized, usually in multiple courses of bleach and alcohol washes, and finally rinsed in sterilized water. This small portion of plant tissue, sometimes only a single cell, is placed on a growth medium, typically containing sucrose as an energy source and one or more plant growth regulators (plant hormones). Usually the medium is thickened with agar to create a gel which supports the explant during growth. Some plants are easily grown on simple media, but others require more complicated media for successful growth; the plant tissue grows and differentiates into new tissues depending on the medium. For example, media containing cytokinin are used to create branched shoots from plant buds.

In vitro culture of plants in a controlled, sterile environment

Multiplication

Multiplication is the taking of tissue samples produced during the first stage and increasing their number. Following the successful introduction and growth of plant tissue, the establishment stage is followed by multiplication. Through repeated cycles of this process, a single explant sample may be increased from one to hundreds and thousands of plants. Depending on the type of tissue grown, multiplication can involve different methods and media. If the plant material grown is callus tissue, it can be placed in a blender and cut into smaller pieces and recultured on the same type of culture medium to grow more callus tissue. If the tissue is grown as small plants called plantlets, hormones are often added that cause the plantlets to produce many small offshoots. After the formation of multiple shoots, these shoots are transferred to rooting medium with a high auxin\ cytokinin ratio. After the development of roots, plantlets can be used for hardening.

Pretransplant

"Hardening" refers to the preparation of the plants for a natural growth environment. Until this stage, the plantlets have been grown in "ideal" conditions, designed to encourage rapid growth. Due to the controlled nature of their maturation, the plantlets often do not have fully functional dermal coverings. This causes them to be highly susceptible to disease and inefficient in their use of water and energy. In vitro conditions are high in humidity, and plants grown under these conditions often do not form a working cuticle and stomata that keep the plant from drying out. When taken out of culture, the plantlets need time to adjust to more natural environmental conditions. Hardening typically involves slowly weaning the plantlets from a high-humidity, low light, warm environment to what would be considered a normal growth environment for the species in question.

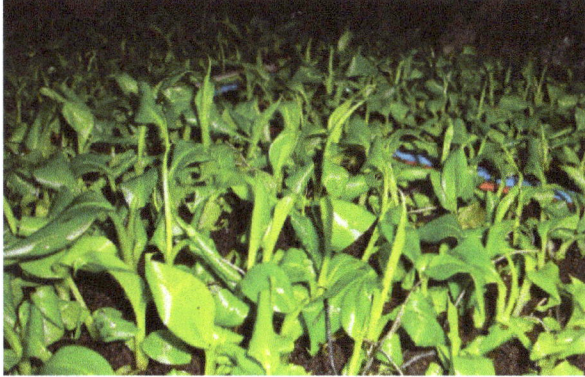

Banana plantlets transferred to soil (with vermicompost) from plant media. This process is done for acclimatization of plantlets to the soil as they were previously grown in plant media. After growing for some days the plantlets are transferred to the field.

This stage involves treating the plantlets/shoots produced to encourage root growth and "hardening." It is performed *in vitro*, or in a sterile "test tube" environment.

Transfer from Culture

Plant tissue cultures being grown at a USDA seed bank, the National Center for Genetic Resources Preservation.

In the final stage of plant micropropagation, the plantlets are removed from the plant media and transferred to soil or (more commonly) potting compost for continued growth by conventional methods.

This stage is often combined with the "pretransplant" stage.

Methods of Micro-propagation

Meristem Culture

In Meristem culture the Meristem and a few subtending leaf primordial are placed into a suitable growing media. An elongated rooted platelet is produced after some weeks, and is transferred to the soil when it has attained a considerable height. A disease free plant can be produced by this method. Experimental result also suggest that this technique can be successfully utilized for rapid multiplication of various herbaceous Plants.

Callus Culture

A callus is mass of undifferentiated parenchymatous cells. When a living plant tissue is placed in an artificial growing medium with other conditions favorable, callus is formed. The growth of callus varies with the homogenous levels of auxin and Cyotkininn and can be manipulated by endogenous supply of these growth regulators in the culture medium. The callus growth and its organogenesis or embryogenesis can be referred into three different stages:

- Stage I: Rapid production of callus after placing the explants in culture medium

- Stage II: The callus is transferred to other medium containing growth regulators for the induction of adventitious organs.

- Stage III: The new plantlet is then exposed gradually to the environmental condition.

Suspension Culture

A cell suspension culture refers to cells and or groups of cells dispersed and growing in an aerated liquid culture medium (Street, 1997, Thorpe1981) is placed in a liquid medium and shaken vigorously and balanced dose of hormones. Suezawa et al. (1988) reported Cyotkininn induced adventitious buds in kiwi fruit in a suspension culture sub- culture for about a week.

Embryo Culture

In embryo culture, the embryo is excised and placed into a culture medium with proper nutrient in aseptic condition. To obtain a quick and optimum growth into plantlets, it is transferred to soil. It is particularly important for the production of interspecific and intergeneric hybrids and to overcome the embryo abortion.

Protoplast Culture

In protoplast culture, the plant cell can be isolated with the help of wall degrading enzymes and growth in a suitable culture medium in a controlled condition for regeneration of plantlets. Under suitable conditions the protoplast develops a cell wall followed by an increase in cell division and differentiation and grows into a new plant. The protoplast are first cultured in liquid medium at 25 to 28 C with a light intensity of 100 to 500 lux or in dark and after undergoing substantial cell division, they are transferred into solid medium congenial or morphogenesis in many horticultural crops response well to protoplast culture.

Advantages

Micropropagation has a number of advantages over traditional plant propagation techniques:

- The main advantage of micropropagation is the production of many plants that are clones of each other.

- Micropropagation can be used to produce disease-free plants.

- It can have an extraordinarily high fecundity rate, producing thousands of propagules while conventional techniques might only produce a fraction of this number.

- It is the only viable method of regenerating genetically modified cells or cells after protoplast fusion.

- It is useful in multiplying plants which produce seeds in uneconomical amounts, or when plants are sterile and do not produce viable seeds or when seed cannot be stored.

- Micropropagation often produces more robust plants, leading to accelerated growth compared to similar plants produced by conventional methods - like seeds or cuttings.

- Some plants with very small seeds, including most orchids, are most reliably grown from seed in sterile culture.

- A greater number of plants can be produced per square meter and the propagules can be stored longer and in a smaller area.

Disadvantages

Micropropagation is not always the perfect means of multiplying plants. Conditions that limits its use include:

- It is very expensive, and can have a labour cost of more than 70%.

- A monoculture is produced after micropropagation, leading to a lack of overall disease resilience, as all progeny plants may be vulnerable to the same infections.

- An infected plant sample can produce infected progeny. This is uncommon as the stock plants are carefully screened and vetted to prevent culturing plants infected with virus or fungus.

- Not all plants can be successfully tissue cultured, often because the proper medium for growth is not known or the plants produce secondary metabolic chemicals that stunt or kill the explant.

- Sometimes plants or cultivars do not come true to type after being tissue cultured. This is often dependent on the type of explant material utilized during the initiation phase or the result of the age of the cell or propagule line.

- Some plants are very difficult to disinfect of fungal organisms.

The major limitation in the use of micropropagation for many plants is the cost of production; for many plants the use of seeds, which are normally disease free and produced in good numbers, readily produce plants in good numbers at a lower cost. For this reason, many plant breeders do not utilize micropropagation because the cost is prohibitive. Other breeders use it to produce stock plants that are then used for seed multiplication.

Mechanisation of the process could reduce labour costs, but has proven difficult to achieve, despite active attempts to develop technological solutions.

Adventitious Shoot Proliferation

Adventitious shoot proliferation in plant cell and tissue culture, in response to hormonal manipulation of the culture medium, require *de novo* differentiation of meristematic region, randomly, all over the tissue other than the pre-existing meristem. It is a multistep process and a series of intracellular events, collectively called induction that occurs before the appearance of morphologically recognizable organs. Micropropagation via adventitious shoot regeneration may occur directly or indirectly via an intervening callus phase. Indirect regeneration often results in somaclonal variations, making this strategy less desirable for large-scale clonal multiplication (Marcotrigiano and Jagannathan 1988; Thorpe et al. 1991). Therefore, regeneration of shoots directly from the explants is regarded as the most reliable method for clonal propagation. Various explants like leaf, cotyledon, embryo and root have been tried with different media combinations by the scientists to obtain adventitious shoot proliferation.

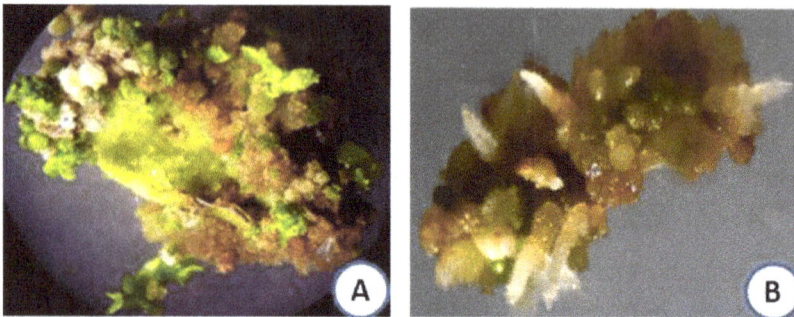

Organogenesis from leaf explants indirectly via callusing
A. Shoot differentiation B. Root differentiation

Organogenic Differentiation

Regeneration of plant from the cultured explant may occur either through differentiation of shoot-buds or somatic embryogenesis.

Direct shoot proliferation from leaf-disc culture

Direct differentiation of somatic embryos from hypocotyl explants

The shoot-bud and embryo formation can be distinguished by the distinct morphological features. The shoot-bud is a monopolar structure. It develops from the procambial strands which establish a connection with the pre-existing vascular tissue dispersed within the callus or the cultured explants.

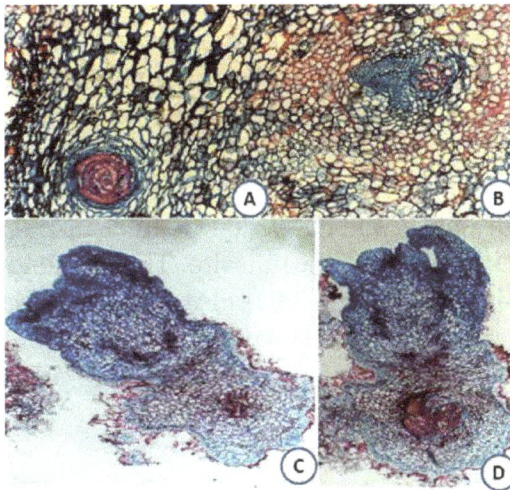

Shoot differentiation from callus tissue. A-B, Development of vascular nodules randomly in the callus, note a small shoot-bud originated from vascular tissue in figure B. C-D, Shoot-buds establish a connection with pre-existing vascular tissue developed from the callus

Plant regeneration from isolated cells, protoplasts or unorganized mass of cells (callus) is generally more difficult than that obtained from the intact explants such as, cotyledons, hypocotyl segments and immature embryos. The regeneration obtained through de novo differentiation of shoot buds or somatic embryogenesis directly from explants may also exhibit genetic variability.

Induction of Organogenic Differentiation

Induction is a multistep process. A schematic representation of adventitious shoot proliferation from leaf-disc culture is presented. A series of intracellular events, collectively called induction, occur before the emergence of morphologically recognizable organs.

A schematic representation of *in vitro* adventitious shoot proliferation from leaf-disc culture

Under the optimal growth regulator combinations, the cells induced to form a specific organ and would continue to develop into that organ even if the inductive growth regulators are removed. Hence, induction favours the irreversible commitment of cells to follow a particular developmental pathway. For example, *Brassica juncea,* undergoes the induction of organogenic differentiation where a cytokinin, BAP induces shoot-bud differentiation at the cut end of the cotyledon petiole. In the absence of BAP (basal medium) only roots are formed at the same site. The cotyledons transferred to basal medium after 11 days of incubation on BAP leads to the development of only shoots and no roots. Similarly, the cotyledons lose the potential to form shoots on BAP medium if they are pre-cultured on BAP free medium for more than 7 days.

Ontogeny of Shoot Buds

Under the optimal conditions, meristems formed from the callus are random and scattered. On transferring to the medium supporting organized growth promotes first the appearance of localized clusters of cambium like cells. These meristemoids (or nodules or growing centers), which may become vascularized due to the appearance of tracheidal cells in the centre, are the site for organ formation in the callus, as seen in above figure. Initially, the meristemoids exhibits plasticity and can form shoots and roots.

Factors Affecting Shoot-bud Differentiation

The genotype and plant growth regulators are well known to affect regeneration frequency. Plant growth regulators play a major role in the regeneration which mainly depends upon the concentration and type of growth regulators used. For *in vitro* differentiation genotype plays equally, if not more critical role as the growth regulator. Besides, there are certain other factors which play a critical role in regeneration are:

Explant

Regenerability of an explant is influenced by several factors such as, the organ from which it is derived, the physiological state of the explant like age of the explant, young vs. mature, position of the explant on the plant and the explant size. Orientation of the explant on the medium and the inoculation density may also affect shoot-bud differentiation. There may be a decline in the number of shoots per culture and the percent cultures showing regeneration with increasing age of the seedlings.

Preparation of Explant

Sharma et al (1990) studied that in cotyledon cultures of *Brassica juncea* , shoot buds or roots are formed at the cut end of the petiole, depending on the culture medium. Lamina lacks this potential. However, the presence of laminar tissue is essential for the petiolar cells to exhibit totipotency. Therefore, the ideal explant to achieve regeneration is the lamina together with a short (1mm) petiole.

Orientation of the Explant

Orientation of explant is proved to be critical for organogenic differentiation in cotyledon cultures of *B. juncea*. Inoculating the cotyledons with their abaxial surface (lower surface away from the stem) in contact with the medium and the petiolar cut end embedded in the medium gave best response. The explants in which due to expansion and curling of the lamina, the petiole lost contact with the medium within 3-5 days after culture, failed to form roots or shoots. Generally, the explants inoculated horizontally on the medium produced three times more shoots than those planted vertically.

Physical Factors

   i.   Explants grown on liquid or semi-solid medium give different degree of organogenesis. In few species, like tobacco, the medium with 1% agar showed only flower formation. With lowering the agar concentration the frequency of flower formation dropped and vegetative bud differentiation occurred. In liquid medium, the tissue exhibited callusing and vegetative bud formation.

ii. The quality of light also influences organogenic differentiation. Alternating light and dark period (diffused light, 15-16 hrs) proved best. Callus maintained under continuous light remained whitish and may not exhibit organogenesis. Blue light promote shoot-bud differentiation whereas red light stimulated rooting in tobacco. Calli of *Brassica oleracea* grown in dark for 20 days formed shoot-buds 12 days after transfer to light while those shifted to light after 12 days of growth in dark differentiated shoots within 9 days.

iii. Skoog (1944) studied the effect of a range of temperature on tobacco callus growth and differentiation. Growth of callus increased with rise in temperature up to 33°C, but for shoot-bud differentiation 18°C was optimum.

Axillary Shoot Proliferation

Axillary buds are usually present in the axil of each leaf and every bud has the potential to develop into a shoot. In nature these buds remain dormant for various periods. The species with strong apical dominance show the growth of axillary buds into shoot only if the terminal bud is removed or injured. The phenomenon of apical dominance is regulated by the interplay of growth regulators. The application of cytokinin to the axillary buds can overcome the apical dominance effect and stimulate the lateral buds to grow rapidly in the presence of terminal buds. If the exogenous growth regulator diminishes the lateral shoot stop growing.

Rate of Shoot Multiplication

In plant tissue culture, the rate of shoot multiplication can be determined by enhanced axillary branching. Due to continuous availability of cytokinin, the shoots formed by the bud, a priori present on the explant (nodal segment or shoot-tip cutting), develops axillary buds which may grow directly into shoots. This process may be repeated several times and the initial explant transformed into a mass of branches. There is a limit to which shoot multiplication can be achieved in a single passage, after which further axillary branching stops. At this stage, if shoots are excised and planted on a fresh medium of same composition, the shoot multiplication cycle can be repeated. This process can go on indefinitely, and can be maintained throughout the year independent of the season and the region.

In some plants, it may not be possible to break apical dominance by the application of growth regulator compositions, and the bud present a priori on the initial explant grows into an unbranched shoot. The rate of shoot multiplication in such cases would depend on the number of nodal cuttings that can be excised from the newly developed shoot at the end of each passage. With this alternative method of enhanced axillary branching, 6-7 fold shoot multiplication every 4-6 weeks could be achieved in the plants with strong apical dominance.

Factors Affecting Axillary Shoot Proliferation

i.   Effects of season on culture establishment

The extent of contamination as well as bud-break is highly dependent on the season. The cultures initiated during spring season (January to April) shows best response not only in terms of the frequency of bud-break but also in the vigor of the shoots with least contamination rate. Since, summer (May-August) is the period that concurs with rainy season in certain regions like India, the cultures are prone to infection. By winter the shoots become old and it is difficult to break the dormant state of the buds.

ii. Effect of carbon source on shoot proliferation

In cultured plant tissues, a continuous supply of carbohydrate from the medium is essential which are needed for growth and organized development of the plant and are necessary as a source of energy and carbon skeletons for biosynthetic process. For shoot induction from axillary buds, three carbon sources, sucrose, glucose and maltose are utilized in maximum plant tissue cultures at a fixed concentration of 30 g l^{-1} . Of these, sucrose is the most commonly used carbohydrate for plant tissue cultures and most culture media have it as the sole carbohydrate source. It favors higher growth of shoot, number of nodes per shoot and the rate of shoot multiplication compare to maltose and glucose. Sucrose is easily recognized and hydrolyzed by cell wall bound invertase into more efficiently utilizable forms of sugars, glucose and fructose which are incorporated into the cells. Glucose, derived from sucrose hydrolysis, is more accessible to the cultured tissues than glucose derived by maltose hydrolysis, due to a rapid sucrose hydrolysis but a slow maltose hydrolysis in the media.

iii. Effect of growth regulators on shoot proliferation

In general, cytokinins favors shoot proliferation and auxins favors root formation. In *S. acmella,* nodal explants bearing two opposite axillary buds were when cultured on MS basal medium or basal medium supplemented with BAP, Kinetin or 2-iP at 3 µM concentration, the frequency of bud-break was appreciable in basal medium but incorporation of BAP to the basal medium has further improved the incidence of bud-break and promoted multiple shoot formation (2 shoots/explant). While the least bud-break was observed on Kinetin supplemented medium and 2-iP was noticed to be inhibitory for axillary bud proliferation. The addition of a low concentration of GA$_3$ to the BAP supplemented medium further promoted multiple shoot formation. On the other hand, single shoot with long internodes was developed from axillary buds in cultures when NAA was added to BAP containing medium. The frequency of bud-break varied with the concentration of the BAP and at its optimum level of 5 µM, 10-fold shoot multiplication occurred every 5 weeks.

Axillary Shoot Proliferation Vs Adventitious Shoot Proliferation

i.   The axillary shoot proliferation is the most popular approach to clonal propagation of crop plants because the cells of the shoot apex are uniformly diploid and are least susceptible to genotypic changes under culture conditions.

ii. Chimeras, whose breakdown is common during adventitious bud proliferation, are perpetuated in shoot-bud culture and, thus, the cause for change in ploidy sometimes. While axillary shoot proliferation favours genetically uniform plant formation.

iii. Moreover adventitious bud formation and callusing methods require denovo differentiation of shoot-buds which is not always possible.

iv. Further, the axillary shoot proliferation is comparatively a quicker method of shoot multiplication as pre-existing meristem only proliferate into shoots, thus, reducing the time required to form de novo meristem formation.

Axillary Shoot Proliferation Vs Conventional Method of Propagation

i.   The conventional method of vegetative propagation by stem cuttings utilizes the ability of axillary buds to take over the function of main shoot in the absence of a terminal bud. However, the number of cuttings that can be taken from a selected plant in a year is extremely limited because in nature the vegetative growth is periodic. In *in vitro* conditions, axillary shoot proliferate irrespective of seasons and regions.

ii. A minimal size of cuttings required in conventional methods is around 24-30 cm in order to establish a plant from it. Thus, it may restrict the multiplication of plants if the stock of parent plant is limited or if the species is endangered.

iii. With axillary shoot proliferation, minimum cutting size required is <1cm, thus, it favours large scale multiplication even with the limited sample.

iv. With the axillary shoot proliferation method, juvenile nodal cuttings are made available throughout the year that helps to maintain faster rate of multiplication compare to conventional methods of vegetative propagation where juvenile phase is short lived and with mature cuttings it is difficult to establish propagation as the buds in the axil under go dormancy.

Experiment

Aim: Clonal propagation by axillary shoot proliferation from nodal explants of *Spilanthes acmella*

Equipments: Autoclave, pH-meter, Magnetic stirrers, Magnetic beads, Weighing balance, Laminar-air-flow

Materials Required : Salts and vitamins of Murashige and Skoog's (MS; 1962) , sucrose, agar, conical flasks, measuring cylinders and beakers of various sizes. Reagent glass bottles for storage, spatula, tissue rolls, distilled water. Cotton plugs, aluminium foils, muslin cloth, scissor, media stocks, 1N NaOH, 1N HCl, myo-inositol. Autoclavable polybags, rubber bands, borosil glass test-tubes (150mm x 25mm without rim). Black markers, micropippete, micropipette-tips, test-tube stands, autoclavable baskets, plastic trays, fresh stem cuttings of *S. acmella* .

Plant Growth regulators (Sigma)

Benzyl amino purine (BAP), 6-furfuryl amino purine (Kinetin), 6-γγ –Dimethylallyl-amino purine (2-iP), Naphthalene acetic acid (NAA), Indole-3-butyric acid.

Protocol

- Nodal segments bearing two opposite axillary buds were collected at monthly intervals to establish *in vitro* cultures.

- Initially, basal medium was examined with three carbon sources *viz.* sucrose, glucose or maltose. Subsequently, modified MS media with major inorganic salts either reduced to half strength (½ MS) or increased to double strength (2 MS) were also tested .

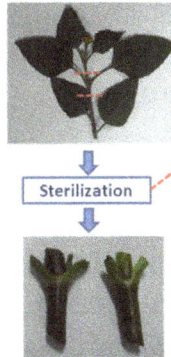

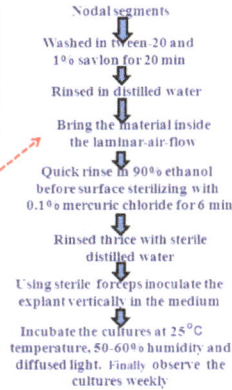

Sterilization

Nodal segments
↓
Washed in tween-20 and 1% savlon for 20 min
↓
Rinsed in distilled water
↓
Bring the material inside the laminar-air-flow
↓
Quick rinse in 90% ethanol before surface sterilizing with 0.1% mercuric chloride for 6 min
↓
Rinsed thrice with sterile distilled water
↓
Using sterile forceps inoculate the explant vertically in the medium
↓
Incubate the cultures at 25°C temperature, 50-60% humidity and diffused light. Finally observe the cultures weekly

Schematic view of establishment of nodal segment cultures of *S. acmella* for clonal propagation

Results

Shoot Multiplication

Nodal explants of *S. acmella* bearing two opposite axillary buds were cultured on MS basal medium or basal medium supplemented with BAP, Kinetin or 2-iP. MS + BAP medium proved optimum for shoot multiplication and results into 10-fold shoot multiplication at every 5 weeks. At the end of the passage each shoot was cut into single node segments and planted on the fresh medium of the same composition. Each node again produced multinodal, multiple shoots after 5 weeks. Shoot multiplication rate in various subculture (S) passages, S_1 =10.2, S_2 =10.3, S_3 =10.4, S_4 =10.6, S_5 =10.6, S_6 =10.5, S_7 =10.5, S_8 =10.5, S_9 =10.6, S_{10} =10.6, was >10 fold from S_1 to S_{10} on MS + BAP (5 µM). Since every time the explants were taken from freshly formed *in vitro* shoots, therefore, no significant difference (variation) was observed in the results.

Nodal segment culture: *invitro* clonal propagation of *Spilanthes acmella* by axillary shoots proliferation

Rooting and Transplantation

Terminal 3-4 cm long portions of shoots after the growth period of 5 weeks are used for rooting. The remaining portions of the shoots can be cut into single node segments and utilized for further multiplication. MS basal medium was tested at full and half (½ MS) strengths of the major inorganic salts and the media are supplemented with 10, 30 and 50 gl^{-1} sucrose. Half MS was distinctly better than full MS basal medium, in terms of length of the shoot, percent rooting, number of roots per shoot and number of laterals present on the roots. The rooting was positively correlated with the sucrose concentration in the medium. On ½ MS + 50 gl^{-1} sucrose, shoots formed more than 35 roots directly from the basal cut end of the shoots. On this medium roots appeared after 2 weeks and maximum response was observed after 4 weeks. Some of the roots had developed laterals.

Rooted shoots were transferred out of culture. The plants were acclimatized by covering the pots with polythene bags to maintain high humidity for 6-7 days and irrigated with major salt solution of MS medium. After 7 days, 3-4 small holes were made in the bag and plantlets were irrigated as before but, at frequent intervals. After 25 days, polythene bags were removed and the acclimatized plants were transferred to a shaded area under natural conditions at a temperature range of 20-25°C with photoperiod of 12/12 h (light/dark). The plantlets were acclimatized successfully. During *in vitro* hardening, shoots elongated, leaves turned greener, and their lamina expanded. Consequently, the plants seemed much healthier and grew more vigorously after *in vitro* hardening.

Nodal segment culture: Rooting and acclimatization

Meristem

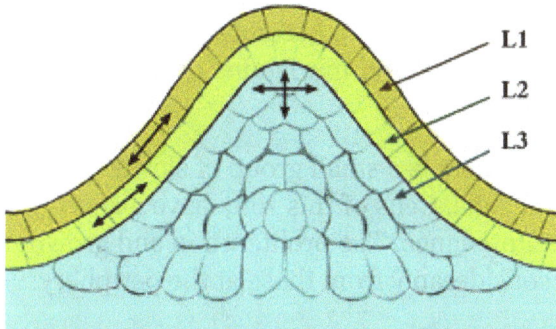

Tunica-Corpus model of the apical meristem (growing tip). The epidermal (L1) and subepidermal (L2) layers form the outer layers called the tunica. The inner L3 layer is called the corpus. Cells in the L1 and L2 layers divide in a sideways fashion, which keeps these layers distinct, whereas the L3 layer divides in a more random fashion.

A meristem is the tissue in most plants containing undifferentiated cells (meristematic cells), found in zones of the plant where growth can take place.

Meristematic cells give rise to various organs of the plant and keep the plant growing. The *shoot apical meristem* (SAM) gives rise to organs like the leaves and flowers, while the *root apical meristem* (RAM) provides the meristematic cells for the future root growth. SAM and RAM cells divide rapidly and are considered indeterminate, in that they do not possess any defined end status. In that sense, the meristematic cells are frequently compared to the stem cells in animals, which have an analogous behavior and function.

The term *meristem* was first used in 1858 by Karl Wilhelm von Nägeli (1817–1891) in his book *Beiträge zur Wissenschaftlichen Botanik* ("Contributions to Scientific Botany"). It is derived from the Greek word *merizein*, meaning to divide, in recognition of its inherent function.

In general, differentiated plant cells cannot divide or produce cells of a different type. Therefore, cell division in the meristem is required to provide new cells for expansion and differentiation of tissues and initiation of new organs, providing the basic structure of the plant body.

Meristematic cells are incompletely or not at all differentiated, and are capable of continued cellular division (youthful). Furthermore, the cells are small and protoplasm fills the cell completely. The vacuoles are extremely small. The cytoplasm does not contain differentiated plastids (chloroplasts or chromoplasts), although they are present in rudimentary form (proplastids). Meristematic cells are packed closely together without intercellular cavities. The cell wall is a very thin *primary cell wall*.

Maintenance of the cells requires a balance between two antagonistic processes: organ initiation and stem cell population renewal.

Apical meristems are the completely undifferentiated (indeterminate) meristems in a plant. These differentiate into three kinds of primary meristems. The primary meristems in turn produce the two secondary meristem types. These secondary meristems are also known as lateral meristems because they are involved in lateral growth.

At the meristem summit, there is a small group of slowly dividing cells, which is commonly called the central zone. Cells of this zone have a stem cell function and are essential for meristem maintenance. The proliferation and growth rates at the meristem summit usually differ considerably from those at the periphery.

Meristems also are induced in the roots of legumes such as soybean, *Lotus japonicus*, pea, and *Medicago truncatula* after infection with soil bacteria commonly called Rhizobium. Cells of the inner or outer cortex in the so-called "window of nodulation" just behind the developing root tip are induced to divide. The critical signal substance is the lipo-oligosaccharide Nod-factor, decorated with side groups to allow specificity of interaction. The Nod factor receptor proteins NFR1 and NFR5 were cloned from several legumes including *Lotus japonicus*, *Medicago truncatula* and soybean (*Glycine max*). Regulation of nodule meristems utilizes long distance regulation commonly called "Autoregulation of Nodulation" (AON). This process involves a leaf-vascular tissue located LRRreceptorkinases (LjHAR1, GmNARK and MtSUNN), CLE peptide signalling, and KAPP interaction, similar to that seen in the CLV1,2,3 system. LjKLAVIER also exhibits a nodule regulation phenotype though it is not yet known how this relates to the other AON receptor kinases.

Apical Meristems

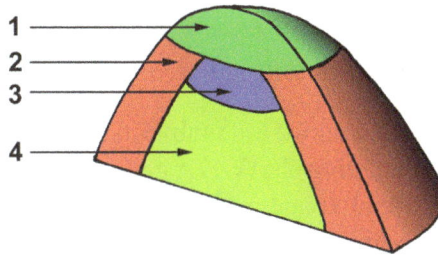

Organisation of an apical meristem (growing tip) 1 - Central zone 2 - Peripheral zone
3 - Medullary (i.e. central) meristem 4 - Medullary tissue

The number of layers varies according to plant type. In general the outermost layer is called the tunica while the innermost layers are the corpus. In monocots, the tunica determine the physical characteristics of the leaf edge and margin. In dicots, layer two of the corpus determine the characteristics of the edge of the leaf. The corpus and tunica play a critical part of the plant physical appearance as all plant cells are formed from the meristems. Apical meristems are found in two locations: the root and the stem. Some Arctic plants have an apical meristem in the lower/middle parts of the plant. It is thought that this kind of meristem evolved because it is advantageous in Arctic conditions.

Shoot Apical Meristems

Shoot apical meristems of *Crassula ovata* (left). Fourteen days later, leaves have developed (right).

The source of all above-ground organs. Cells at the shoot apical meristem summit serve as stem cells to the surrounding peripheral region, where they proliferate rapidly and are incorporated into differentiating leaf or flower primordia.

The shoot apical meristem is the site of most of the embryogenesis in flowering plants. Primordia of leaves, sepals, petals, stamens and ovaries are initiated here at the rate of one every time interval, called a plastochron. It is where the first indications that flower development has been evoked are manifested. One of these indications might be the loss of apical dominance and the release of otherwise dormant cells to develop as auxiliary shoot meristems, in some species in axils of primordia as close as two or three away from the apical dome. The shoot apical meristem consists of 4 distinct cell groups:

- Stem cells

- The immediate daughter cells of the stem cells

- A subjacent organising centre

- Founder cells for organ initiation in surrounding regions

The four distinct zones mentioned above are maintained by a complex signalling pathway. In *Arabidopsis thaliana*, 3 interacting *CLAVATA* genes are required to regulate the size of the stem cell reservoir in the shoot apical meristem by controlling the rate of cell division. CLV1 and CLV2 are predicted to form a receptor complex (of the LRR receptor-like kinase family) to which CLV3 is a ligand. CLV3 shares some homology with the ESR proteins of maize, with a short 14 amino acid region being conserved between the proteins. Proteins that contain these conserved regions have been grouped into the CLE family of proteins.

CLV1 has been shown to interact with several cytoplasmic proteins that are most likely involved in downstream signalling. For example, the CLV complex has been found to be associated with Rho/Rac small GTPase-related proteins. These proteins may act as an intermediate between the CLV complex and a mitogen-activated protein kinase (MAPK), which is often involved in signalling cascades. KAPP is a kinase-associated protein phosphatase that has been shown to interact with CLV1. KAPP is thought to act as a negative regulator of CLV1 by dephosphorylating it.

Another important gene in plant meristem maintenance is *WUSCHEL* (shortened to *WUS*), which is a target of CLV signalling. *WUS* is expressed in the cells below the stem cells of the meristem and its presence prevents the differentiation of the stem cells. CLV1 acts to promote cellular differentiation by repressing *WUS* activity outside of the central zone containing the stem cells. *SHOOT MERISTEMLESS* (*STM*) also acts to prevent the differentiation of stem cells by repressing the expression of MYB genes that are involved in cellular differentiation.

Root Apical Meristem

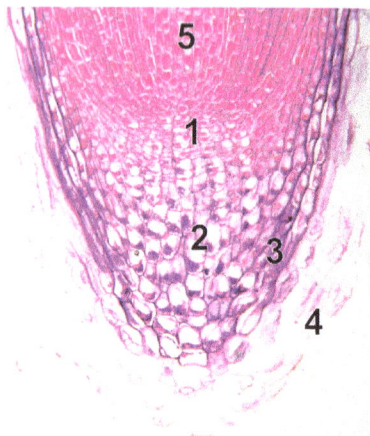

10x microscope image of root tip with meristem 1 - quiescent center 2 - calyptrogen (live rootcap cells) 3 - rootcap 4 - sloughed off dead rootcap cells 5 - procambium

Unlike the shoot apical meristem, the root apical meristem produces cells in two dimensions. It harbors two pools of stem cells around an organizing center called the quiescent center (QC) cells and together produce most of the cells in an adult root. At its apex, the root meristem is covered by the root cap, which protects and guides its growth trajectory. Cells are continuously sloughed off the outer surface of the root cap. The QC cells are characterized by their low mitotic activity. Evidence suggests that the QC maintains the surrounding stem cells by preventing their differentiation, via signal(s) that are yet to be discovered. This allows a constant supply of new cells in the meristem required for continuous root growth. Recent findings indicate that QC can also act as a reservoir of stem cells to replenish whatever is lost or damaged. Root apical meristem and tissue patterns become established in the embryo in the case of the primary root, and in the new lateral root primordium in the case of secondary roots.

Intercalary Meristem

In angiosperms, intercalary meristems occur only in monocot (in particular, grass) stems at the base of nodes and leaf blades. Horsetails also exhibit intercalary growth. Intercalary meristems are capable of cell division, and they allow for rapid growth and regrowth of many monocots. Intercalary meristems at the nodes of bamboo allow for rapid stem elongation, while those at the base of most grass leaf blades allow damaged leaves to rapidly regrow. This leaf regrowth in grasses evolved in response to damage by grazing herbivores, but is more familiar to us in response to lawnmowers.

Floral Meristem

When plants begin the developmental process known as flowering, the shoot apical meristem is transformed into an inflorescence meristem, which goes on to produce the floral meristem, which produces the sepals, petals, stamens, and carpels of the flower.

In contrast to vegetative apical meristems and some exflorescence meristems, floral meristems cannot continue to grow indefinitely. Their future growth is limited to the flower with a particular size and form. The transition from shoot meristem to floral meristem requires floral meristem identity genes, that both specify the floral organs and cause the termination of the production of stem cells. *AGAMOUS (AG)* is a floral homeotic gene required for floral meristem termination and necessary for proper development of the stamens and carpels. *AG* is necessary to prevent the conversion of floral meristems to inflorescence shoot meristems, but is not involved in the transition from shoot to floral meristem. AG is turned on by the floral meristem identity gene *LEAFY (LFY)* and *WUS* and is restricted to the centre of the floral meristem or the inner two whorls. This way floral identity and region specificity is achieved. WUS activates AG by binding to a consensus sequence in the AG's second intron and LFY binds to adjacent recognition sites. Once AG is activated it represses expression of WUS leading to the termination of the meristem.

Through the years, scientists have manipulated floral meristems for economic reasons. An example is the mutant tobacco plant "Maryland Mammoth." In 1936, the department of agriculture of Switzerland performed several scientific tests with this plant. "Maryland Mammoth" is peculiar in that it grows much faster than other tobacco plants.

Apical Dominance

Apical dominance is phenomenon where one meristem prevents or inhibits the growth of other meristems. As a result, the plant will have one clearly defined main trunk. For example, in trees, the tip of the main trunk bears the dominant meristem. Therefore, the tip of the trunk grows rapidly and is not shadowed by branches. If the dominant meristem is cut off, one or more branch tips will assume dominance. The branch will start growing faster and the new growth will be vertical. Over the years, the branch may begin to look more and more like an extension of the main trunk. Often several branches will exhibit this behaviour after the removal of apical meristem, leading to a bushy growth.

The mechanism of apical dominance is based on the plant hormone auxin. It is produced in the apical meristem and transported towards the roots in the cambium. If apical dominance is complete, it prevents any branches from forming as long as the apical meristem is active. If the dominance is incomplete, side branches will develop.

Recent investigations into apical dominance and the control of branching have revealed a new plant hormone family termed strigolactones. These compounds were previously known to be involved in seed germination and communication with mycorrhizal fungi and are now shown to be involved in inhibition of branching.

Diversity in Meristem Architectures

Is the mechanism of being *indeterminate* conserved in the SAMs of the plant world? The SAM contains a population of stem cells that also produce the lateral meristems while the stem elongates. It turns out that the mechanism of regulation of the stem cell number might indeed be evolutionarily conserved. The *CLAVATA* gene *CLV2* responsible for maintaining the stem cell population in *Arabidopsis thaliana* is very closely related to the Maize gene *FASCIATED EAR 2(FEA2)* also involved in the same function. Similarly, in Rice, the *FON1-FON2* system seems to bear a close relationship with the CLV signaling system in *Arabidopsis thaliana*. These studies suggest that the regulation of stem cell number, identity and differentiation might be an evolutionarily conserved mechanism in monocots, if not in angiosperms. Rice also contains another genetic system distinct from *FON1-FON2*, that is involved in regulating stem cell number. This example underlines the innovation that goes about in the living world all the time.

Role of the KNOX-family Genes

Note the long spur of the above flower. Spurs attract pollinators and confer pollinator specificity. *(Flower:Linaria dalmatica)*

Complex leaves of *C. hirsuta* are a result of KNOX gene expression

Genetic screens have identified genes belonging to the KNOX family in this function. These genes essentially maintain the stem cells in an undifferentiated state. The KNOX family has undergone quite a bit of evolutionary diversification, while keeping the overall mechanism more or less similar. Members of the KNOX family have been found in plants as diverse as Arabidopsis thaliana, rice, barley and tomato. KNOX-like genes are also present in some algae, mosses, ferns and gymnosperms. Misexpression of these genes leads to formation of interesting morphological features. For example, among members of *Antirrhinae,* only the species of genus Antirrhinum lack a structure called spur in the floral region. A spur is considered an evolutionary innovation because it defines pollinator specificity and attraction. Researchers carried out transposon mutagenesis in *Antirrhinum majus,* and saw that some insertions led to formation of spurs that were very similar to the other members of *Antirrhinae,* indicating that the loss of spur in wild *Antirrhinum majus* populations could probably be an evolutionary innovation.

The KNOX family has also been implicated in leaf shape evolution. One study looked at the pattern of KNOX gene expression in *A. thaliana,* that has simple leaves and *Cardamine hirsuta*, a plant having complex leaves. In *A. thaliana*, the KNOX genes are completely turned off in leaves, but in *C.hirsuta*, the expression continued, generating complex leaves. Also, it has been proposed that the mechanism of KNOX gene action is conserved across all vascular plants, because there is a tight correlation between KNOX expression and a complex leaf morphology.

Primary Meristems

Apical meristems may differentiate into three kinds of primary meristem:

- Protoderm: lies around the outside of the stem and develops into the epidermis.

- Procambium: lies just inside of the protoderm and develops into primary xylem and primary phloem. It also produces the vascular cambium, and cork cambium, secondary meristems. The cork cambium further differentiates into the phelloderm (to the inside) and the phellem, or cork (to the outside). All three of these layers (cork cambium, phellem and phelloderm) constitute the periderm. In roots, the procambium can also give rise to the pericycle, which produces lateral roots in eudicots.

- Ground meristem: develops into the cortex and the pith. Composed of parenchyma, collenchyma and sclerenchyma cells.

These meristems are responsible for primary growth, or an increase in length or height, which were discovered by scientist Joseph D. Carr of North Carolina in 1943.

Secondary Meristems

There are two types of secondary meristems, these are also called the *lateral meristems* because they surround the established stem of a plant and cause it to grow laterally (i.e., larger in diameter).

- Vascular cambium, which produces secondary xylem and secondary phloem. This is a process that may continue throughout the life of the plant. This is what gives rise to wood in plants. Such plants are called arborescent. This does not occur in plants that do not go through secondary growth (known as herbaceous plants).

- Cork cambium, which gives rise to the periderm, which replaces the epidermis.

Indeterminate Growth of Meristems

Though each plant grows according to a certain set of rules, each new root and shoot meristem can go on growing for as long as it is alive. In many plants, meristematic growth is potentially indeterminate, making the overall shape of the plant not determinate in advance. This is the primary growth. Primary growth leads to lengthening of the plant body and organ formation. All plant organs arise ultimately from cell divisions in the apical meristems, followed by cell expansion and differentiation. Primary growth gives rise to the apical part of many plants.

The growth of nitrogen fixing nodules on legume plants such as soybean and pea is either determinate or indeterminate. Thus, soybean (or bean and Lotus japonicus) produce determinate nodules (spherical), with a branched vascular system surrounding the central infected zone. Often, Rhizobium infected cells have only small vacuoles. In contrast, nodules on pea, clovers, and Medicago truncatula are indeterminate, to maintain (at least for some time) an active meristem that yields new cells for Rhizobium infection. Thus zones of maturity exist in the nodule. Infected cells usually possess a large vacuole. The plant vascular system is branched and peripheral.

Cloning

Under appropriate conditions, each shoot meristem can develop into a complete new plant or clone. Such new plants can be grown from shoot cuttings that contain an apical meristem. Root apical meristems are not readily cloned, however. This cloning is called asexual reproduction or vegetative reproduction and is widely practiced in horticulture to mass-produce plants of a desirable genotype. This process is also known as meri-cloning.

Propagating through cuttings is another form of vegetative propagation that initiates root or shoot production from secondary meristematic cambial cells. This explains why basal 'wounding' of shoot-borne cuttings often aids root formation.

Meristem Culture For Virus Elimination

The plants infected with bacteria and fungi can be treated by bactericidal and fungicidal compounds, there is no commercially available treatment to cure virus-infected plants. A large numbers of viruses are not transmitted through seeds. Therefore, it would be possible to obtain virus free plants from infected individuals by using seeds as propagules. However, genetic variation often occurs from the sexually reproduced plants when propagated by seeds. Generally, clonal multiplication of cultivars can be achieved by vegetative propagation. However, where the entire population of the clone is infected the only way to obtain pathogen-free stock is to eradicate the pathogen from vegetative parts of the plants and regenerate full plants from such tissues. Once pathogen free plants are obtained they can be multiplied indefinitely under conditions which would protect them from chance reinfection.

In Vitro Meristem and Shoot-tip Culture

Explant Terminology

The scientist who have attempted to recover pathogen-free plants through tissue culture techniques have indiscriminately designated the explants required to initiate cultures as 'shoot-tip', 'tip-meristem' and 'meristem-tip'. The apical meristem of a shoot is the portion lying distal to the youngest leaf primordium, it measures up to about 100μm in diameter and 250μm in length. The apical meristem together with one to three young leaf primordia, measuring 100-500μm, constitutes the shoot-apex. Although the chances of eradicating viruses is higher through 'meristem' culture, in most successful reports virus-free plants have been raised by culturing 100-1000μm long explants which could be according to the above definition is referred as 'shoot-tip'. To distinguish it from the *in vivo* technique of propagation through shoot-tip cuttings, the term 'meristem-tip' culture has been preferred for *in vitro* culture of small shoot-tips.

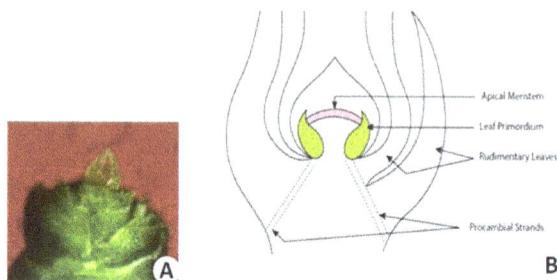

A shoot-tip explant; **B.** A cross section of shoot-tip meristem

Reasons for Escape of Meristem from Virus

It is well known that the distribution of viruses in plants is uneven. In infected plants the apical meristems are generally either free or carry a very low concentration of the viruses. In older tissues the virus titer increases with increasing distance from the meristem-tips. The reasons proposed for the escape of meristem from virus invasion are:

(a) Viruses readily move in a plant body through the vascular system which is absent in the meristem.

(b) The alternative method of cell-to-cell movement of the virus through plasmodesmata is rather too slow to keep pace with the actively growing tip.

(c) High metabolic activity in the actively dividing meristem cells does not allow virus replication.

(d) The 'virus inactivating systems' in the plant body, if any, has higher activity in the meristem than in any other region. Thus, the meristem is protected from infection.

(e) A high endogenous auxin level in shoot apices may inhibit virus multiplication.

Factors Affecting Eradication of Virus Through Meristem Tip Culture

Culture medium, explant size and incubation conditions affecting plant regeneration from meristem-tip cultures have pronounced effect on virus eradication. Besides, thermotherapy or chemotherapy and physiological stage of the explants also affect virus elimination by shoot-tip culture.

Culture Medium

The nutrients, growth regulators and nature of the medium highly influence the development of virus free plants from meristem tip cultures. Maximum success is achieved from Murashige & Skoog's (MS) medium which promoted healthy, green shoot development compare to other nutrient media. The main reason for the suitability of medium for meristem-tip culture could be the presence of high levels of K^+ and NH_4^+ ions. There is no critical assessment on the role of various vitamins or amino acids but sucrose or glucose is the most commonly used carbon source in the medium, at the range of 2-4%, to raise virus free plants from meristem-tip cultures.

Large meristem-tip explants, measuring 500µm or more in length, may give rise to plants even in the basal medium but generally the presence of an auxin or a cytokinin or both plays a major role in the development of excised apical meristem. In angiosperms, the meristematic dome in the shoot-tip does not synthesize auxin on its own, but it is supplied by the second pair of youngest leaf primordia. Therefore, for development of excised meristem in culture, without the leaf primordia, requires the supply of exogenous auxin. The plants requiring only auxin must have a high endogenous cytokinin level in their meristems. Among auxins, the use of 2,4-D should be avoided which promotes only callusing. NAA and IAA are widely used auxins and NAA being preferred due to better stability. The role of GA_3 is also emphasized by few authors which is suggested to promote better growth and differentiation and suppresses callusing from meristem explants. Both liquid and semi-solid media have been tried for meristem–tip culture but, agar medium is generally preferred.

Explant Size

The survival of the meristem tips, under the controlled condition, is determined by the size of the explant. The larger the explant, the greater are the chances of plant regeneration. However, the survival of the explants can not be treated independent of the efficiency with which virus elimination is achieved that is inversely related to the size of the explant. Thus, explants should be small enough to eradicate viruses and large enough to be able to develop into a complete plant. Besides the size of the explant, the presence of leaf primordia influences the ability of the meristems to form plants. In some plants it is essential to excise shoot meristems with two to three leaf primordia. Smith and Murashige (1970) have suggested that leaf primordia supply auxin and cytokinin to the meristem necessary for its growth and differentiation. In a culture medium containing essential growth regulators, the excised meristems domes develop bipolar axes very quickly during reorganization. Once the root-shoot axis is established further development follows the same pattern as that of seedlings.

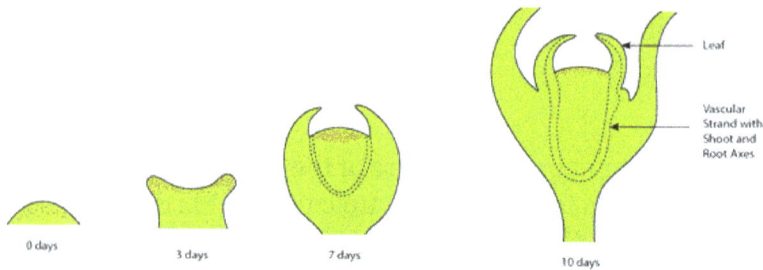

Schematic representation of development of bipolar axes by meristem culture

Storage Conditions

Generally, light incubation of meristem tip culture is found better than dark incubation. The light intensity could range from 100 lx to 4000 lx which should increase in succession as the differentiation of meristem explant progresses. There are no clear information on the effect of temperature on regeneration of plants from excised meristem tips. The cultures are normally stored under room temperature (25±2°C) conditions.

Physiological Conditions of the Explant

Meristem-tips should be collected from actively growing buds. In few cases, the tips taken from terminal buds proved better than those taken from axillary buds. Seeing the higher number of axillary buds present per shoot, in majority of the reports axillary buds were utilized as explants to increase the overall production of virus-free plants. The time excision of buds is also critical, specially for the trees with periodic growth. For example, in temperate trees the growth of the plant is limited to only a very short period in the spring and afterwards dormancy starts. In such cases, the meristem-tip cultures can be raised during spring only for increased success rate.

Thermotherapy

Often, apical meristems are not always free of virus and it can't be considered as a universal occurrence. There are certain viruses like, Tobacco Mosaic Virus (TMV), Potato Virus X (PVX) and Cucumber Mosaic Virus (CMV), which invade the meristematic region of the growing tips and interrupts the growth of the meristematic tissue. In such cases also it has been possible to obtain virus-free plants by combining meristem-tip culture with thermotherapy. In this technique, first the mother plants are exposed to heat treatment before excising the meristem-tips or, alternatively, shoot-tip cultures are exposed to high temperature regimes (35°C-40°C) for certain duration (6h to 6 weeks) to obtain virus free plants. In the later case, continuous exposure to very high temperature causes deterioration of the host tissues. The first procedure of treating the mother plant has added advantage where larger explants can be taken from the treated stock and thus, favors relatively higher chances of the tip survival.

Chemotherapy

Chemotherapy is the treatment of an ailment by chemicals especially by killing micro-organisms. It will not eradicate the virus completely. However, a large number of antibiotics, growth regulators, amino acids, purines and pyrimidines can be tested for inactivation of viruses. A nucleotide analogue ribavirin has been found to be the most efficient viracide for plant viruses. This broad spectrum antiviral agent, effective against both plant and animal, was reported to eliminate PVY, CMV and TMV from tobacco explant cultures, Chlorotic Leaf Spot Virus (CLSV) in apple cultures when incorporated into the medium. Vidarabine (adenine arabinoside) and antiserum are also known to reduce the titre of viruses. The effectivity of the compound may vary with the virus and the host genotype.

Virus Elimination Through Callus Culture

It is a general observation that not all the cells in a calli uniformly carry the pathogen when raised from infected tissues. The two possible reasons for the escape of some cells of a systematically infected callus from virus infection are: (a) virus replication is unable to keep pace with cell proliferation, and (b) some cells acquire resistance to virus infection through mutagenesis. Therefore, it is possible to raise virus-free plants from infected shoot-tip calli. However, genetic instability of cultured cells and lack of plant regeneration in callus cultures of some plants poses the limitations of using calli for virus elimination.

Virus Indexing

Even after subjecting the meristem-tips to various treatments favoring virus eradication, only a proportion of the cultures yield virus free plants. Therefore, it is required to test all plants, regenerated through meristem-tip or callus cultures, for specific viruses before being used as mother plant to produce virus-free stock. The individual plants consistently showing negative results for virus titre can be marked as 'virus tested' for specific virus/es and can be released for commercial purposes. The following tests can be performed for virus testing:

i.   The simplest test for the presence or absence of viruses in plant tissues is to examine the leaves and stem for the visible symptoms characteristic of the virus.

ii. Another test is the sap transmission test or 'bioassay test' or 'infectivity test'. It is a very sensitive test and can be performed at a commercial scale. To perform this, ground the test leaves in equal volume (w/v) of 0.5M phosphate buffer using a mortar and pestle. Leaves of the indicator plant (a plant very susceptible to specific viruses), dusted with 600-grade carborundum, are swabbed with the leaf sap from the test plant. After 5 min the incubated leaves are gently washed with water to remove the residual inoculum. The inoculated indicator plants are maintained in a glasshouse, separate from other plants. It may take several days to several weeks, depending on the nature of virus and the virus titre, for the symptoms to appear on the indicator plants. It is used to detect some viruses and viroids but is a slow process requiring several days to months.

iii. The third method, enzyme-linked immunosorbant assay (ELISA), is more rapid serological test which allows quick detection of important viruses. It relies on the use of antibodies prepared against the viral coat protein, requires only a small amount of antiserum and can be performed with simple equipment. However, it is not applicable to viroids and viruses which have lost their coat proteins

Plant Embryogenesis

Plant embryogenesis is the process that produces a plant embryo from a fertilized ovule by asymmetric cell division and the differentiation of undifferentiated cells into tissues and organs. It occurs during seed development, when the single-celled zygote undergoes a programmed pattern of cell division resulting in a mature embryo. A similar process continues during the plant's life within the meristems of the stems and roots.

Seeds

Embryogenesis occurs naturally as a result of sexual fertilization and the formation of the *zygotic embryos*. The embryo along with other cells from the mother plant develops into the seed or the next generation, which, after germination, grows into a new plant.

Embryogenesis may be divided up into two phases, the first involves morphogenetic events which form the basic cellular pattern for the development of the shoot-root body and the primary tissue layers; it also programs the regions of meristematic tissue formation. The second phase, or postembryonic development, involves the maturation of cells, which involves cell growth and the storage of macromolecules (such as oils, starches and proteins) required as a 'food and energy supply' during germination and seedling growth. Embryogenesis involves cell growth and division, cell differentiation and programmed cellular death. The zygotic embryo is formed following double fertilisation of the ovule, giving rise to two distinct structures: the plant embryo and the endosperm which together go on to develop into a seed. Seeds may also develop without fertilization, which is referred to as apomixis. Plant cells can also be induced to form embryos in plant tissue culture; such embryos are called *somatic embryos* and can be used to generate new plants from single cells.

Following fertilization, the zygote undergoes an asymmetrical cell division that gives rise to a small apical cell, which becomes the embryo and a large basal cell (called the suspensor), which functions to provide nutrients from the endosperm to the growing embryo, and the product is an ovoid mass of tissue called the proembryo. From the eight cell stage (octant stage) onwards, the zygotic embryo shows clear embryo patterning, which forms the main axis of polarity, and the linear formation of future structures. These structures include the shoot meristem, cotyledons, hypocotyl, and the root and root meristem: they arise from specific groups of cells as the young embryo divides and their formation has been shown to be position-dependent.

In the globular stage, the embryo develops radial patterning through a series of cell divisions, with the outer layer of cells differentiating into the 'protoderm.' The globular embryo can be thought of as two layers of inner cells with distinct developmental fates; the apical layer will go on to produce cotyledons and shoot meristem, while the lower layer produces the hypocotyl and root meristem. Bilateral symmetry is apparent from the heart stage; provascular cells will also differentiate at this stage. In the subsequent torpedo and cotyledonary stages of embryogenesis, the embryo completes its growth by elongating and enlarging.

In a dicot embryo, the hypophysis, which is the uppermost cell of the suspensor, differentiates to form part of the root cap called Columella.

Plant Growth and Buds

Embryonic tissue is made up of actively growing cells and the term is normally used to describe the early formation of tissue in the first stages of growth. It can refer to different stages of the sporophyte and gametophyte plant; including the growth of embryos in seedlings, and to meristematic tissues, which are in a persistently embryonic state, to the growth of new buds on stems.

In both gymnosperms and angiosperms, the young plant contained in the seed, begins as a developing egg-cell formed after fertilization (sometimes without fertilization in a process called apomixis) and becomes a plant embryo. This embryonic condition also occurs in the buds that form on stems. The buds have tissue that has differentiated but not grown into complete structures. They can be in a resting state, lying dormant over winter or when conditions are dry, and then commence growth when conditions become suitable. Before they start growing into stem, leaves, or flowers, the buds are said to be in an embryonic state.

Somatic Embryogenesis

Somatic embryos are formed from plant cells that are not normally involved in the development of embryos, i.e. ordinary plant tissue. No endosperm or seed coat is formed around a somatic embryo. Applications of this process include: clonal propagation of genetically uniform plant material; elimination of viruses; provision of source tissue for genetic transformation; generation of whole plants from single cells called protoplasts; development of synthetic seed technology. Cells derived from competent source tissue are cultured to form an undifferentiated mass of cells called a callus. Plant growth regulators in the tissue culture medium can be manipulated to induce callus formation and subsequently changed to induce embryos to form the callus. The ratio of different plant growth regulators required to induce callus or embryo formation varies with the type of plant. Asymmetrical cell division also seems to be important in the development of somatic embryos, and while failure to form the suspensor cell is lethal to zygotic embryos, it is not lethal for somatic embryos.

Somatic Embryogenesis

Somatic embryogenesis is an artificial process in which a plant or embryo is derived from a single somatic cell or group of somatic cells. Somatic embryos are formed from plant cells that are not normally involved in the development of embryos, i.e. ordinary plant tissue. No endosperm or seed coat is formed around a somatic embryo. Applications of this process include: clonal propagation of genetically uniform plant material; elimination of viruses; provision of source tissue for genetic transformation; generation of whole plants from single cells called protoplasts; development of synthetic seed technology. Cells derived from competent source tissue are cultured to form an undifferentiated mass of cells called a callus. Plant growth regulators in the tissue culture medium can be manipulated to induce callus formation and subsequently changed to induce embryos to form from the callus. The ratio of different plant growth regulators required to induce callus or embryo formation varies with the type of plant. Somatic embryos are mainly produced *in vitro* and for laboratory purposes, using either solid or liquid nutrient media which contain plant growth regulators (PGR's). The main PGRs used are auxins but can contain cytokinin in a smaller amount. Shoots and roots are monopolar while somatic embryos are bipolar, allowing them to form a whole plant without culturing on multiple media types. Somatic embryogenesis has served as a model to understand the physiological and biochemical events that occur during plant developmental processes as well as a component to biotechnological advancement. The first documentation of somatic embryogenesis was by Steward et al. in 1958 and Reinert in 1959 with carrot cell suspension cultures.

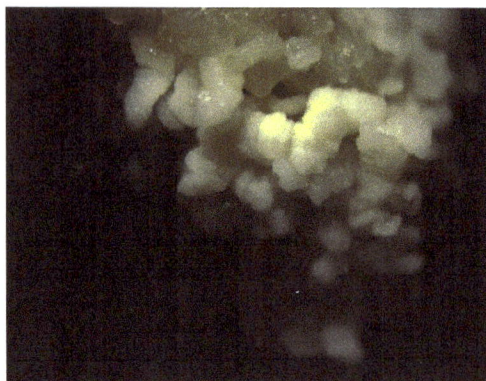

Switchgrass somatic embryos

Direct and Indirect Embryogenesis

Somatic embryogenesis has been described to occur in two ways: directly or indirectly. Direct embryogenesis occurs when embryos are started directly from explant tissue creating an identical clone. Indirect embryogenesis occurs when explants produced undifferentiated, or partially differentiated, cells (often referred to as callus) which then is maintained or differentiated into plant tissues such as leaf, stem, or roots.

Plant Regeneration by Somatic Embryogenesis

Plant regeneration via somatic embryogenesis occurs in five steps: initiation of embryogenic cultures, proliferation of embryogenic cultures, prematuration of somatic embryos, maturation of somatic embryos and plant development on nonspecific media. Initiation and proliferation occur on a medium rich in auxin, which induces differentiation of localized meristematic cells. The auxin typically used is 2,4-D. Once transferred to a medium with low or no auxin, these cells can then develop into mature embryos. Germination of the somatic embryo can only occur when it is mature enough to have functional root and shoot apices

Factors influencing Somatic Embryogenesis

Factors and mechanisms controlling cell differentiation in somatic embryos are relatively ambiguous. Certain compounds excreted by plant tissue cultures and found in culture media have been shown necessary to coordinate cell division and morphological changes. These compounds have been identified by Chung et al. as various polysaccharides, amino acids, growth regulators, vitamins, low molecular weight compounds and polypeptides. Several signaling molecules known to influence or control the formation of somatic embryos have been found and include extracellular proteins, arabinogalactan proteins and lipochitooligosaccharides. Temperature and lighting can also affect the maturation of the somatic embryo.

Uses of Somatic Embryogenesis

- Plant transformations

- Mass propagation

Forestry Related Example

The development of somatic embryogenesis procedures has given rise to research on seed storage proteins (SSPs) of woody plants for tree species of commercial importance, i.e., mainly gymnosperms, including white spruce. In this area of study, SSPs are used as markers to determine the embryogenic potential and competency of the embryogenic system to produce a somatic embryo biochemically similar to its zygotic counterpart (Flinn et al. 1991, Beardmore et al. 1997).

Grossnickle et al. (1992) compared interior spruce seedlings with emblings during nursery development and through a stock quality assessment program immediately before field outplanting. Seedling shoot height, root collar diameter, and dry weight increased at a greater rate in seedlings than in emblings during the first half of the first growing season, but thereafter shoot growth was similar among all plants. By the end of the growing season, seedlings were 70% taller than emblings, had greater root collar diameter, and greater shoot dry weight. Root dry weight increased more rapidly in seedlings than in emblings during the early growing season

During fall acclimation, the pattern of increasing dormancy release index and increasing tolerance to freezing was similar in both seedlings and emblings. Root growth capacity decreased then increased during fall acclimation, with the increase being greater in seedlings.

Assessment of stock quality just prior to planting showed that: emblings had greater water use efficiency with decreasing predawn shoot water potential compared with seedlings; seedlings and emblings had similar water movement capability at both high and low root temperatures; net photosynthesis and needle conductance at low root temperatures were greater in seedlings than in emblings; and seedlings had greater root growth than emblings at 22°C root, but root growth among all plants was low at 7.5°C root temperature.

Growth and survival of interior spruce 313B Styroblock® seedlings and emblings after outplanting on a reforestation site were determined by Grossnickle and Major (1992). For both seedlings and emblings, osmotic potential at saturation (ψ_{sat}) and turgor loss point (ψ_{tip}) increased from a low of -1.82 and -2.22 MPa, respectively, just prior to planting to a seasonal high of -1.09 and -1.21 MPa, respectively, during active shoot elongation. Thereafter, seedlings and emblings (ψ_{sat}) and (ψ_{tip}) declined to -2.00 and -2.45 MPa, respectively, at the end of the growing season, which coincided with the steady decline in site temperatures and a cessation of height growth. In general, seedlings and emblings had similar ψ_{sat} and ψ_{tip} values through the growing season, and also had similar shifts in seasonal patterns of maximum modulus of elasticity, sympalstic fraction, and relative water content at turgor loss point.

Grossnickle and Major (1992) found that year-old and current-year needles of both seedlings and emblings had a similar decline in needle conductance with increasing vapour pressure deficit. Response surface models of current-year needles net photosynthesis (P_n) response to vapour pressure deficit (VPD) and photosynthetically active radiation (PAR) showed that emblings had 15% greater P_n at VPD of less than 3.0 kPa and PAR greater than 1000 μmol m^{-2}s^{-1}. Year-old and current-year needles of seedlings and emblings showed similar patterns of water use efficiency.

Rates of shoot growth in seedlings and emblings through the growing season were also similar to one another. Seedlings had larger shoot systems both at the time of planting and at the end of the growing season. Seedlings also had greater root development than emblings through the growing season, but root:shoot ratios for the 2 stock types were similar at the end of the growing season, when the survival rates for seedlings and emblings were 96% and 99%, respectively.

Problems Associated with Somatic Embryogenesis

- High chance of mutations

- Difficult method

- Loss of regenerative ability

- High percentage of albino shoots during regeneration

- Not possible with all plant species and must be optimized for each species and its use

Tracking and Fate Maps

Understanding the formation of a somatic embryo through establishment of morphological and molecular markers is important for construction of a fate map. The fate map is the foundation in which to build further research and experimentation. Two methods exist to construct a fate map: synchronous cell-division and time-lapse tracking. The latter typically works more consistently because of cell-cycle-altering chemicals and centrifuging involved in synchronous cell-division.

Angiosperms

Embryo development in angiosperms is divided into several steps. The zygote is divided asymmetrically forming a small apical cell and large basal cell. The organizational pattern is formed in the globular stage and the embryo then transitions to the cotyledonary stage. Embryo development differs in monocots and dicots. Dicots pass through the globular, heart-shaped, and torpedo stages while monocots pass through globular, scuetellar, and coleoptilar stages.

Many culture systems induce and maintain somatic embryogenesis by continuous exposure to 2,4-dichlorophenoxyacetic acid. Abscisic acid has been reported to induce somatic embryogenesis in seedlings. After callus formation, culturing on a low auxin or hormone free media will promote somatic embryo growth and root formation. In monocots, embryogenic capability is usually restricted to tissues with embryogenic or meristematic origin. Somatic cells of monocots differentiate quickly and then lose mitotic and morphogenic capability. Differences of auxin sensitivity in embryogenic callus growth between different genotypes of the same species show how variable auxin responses can be.

Carrot *Daucus carota* was the first and most understood species with regard to developmental pathways and molecular mechanisms. Time-lapse tracking by Toonen et al.

(1994) showed that morphology of competent cells can vary based on shape and cytoplasm density. Five types of cells were identified from embryonic suspension: spherical cytoplasm-rich, spherical vacuolated, oval vacuolated, elongated vacuolated, and irregular shaped cells. Each type of cell multiplied with certain geometric symmetry. They developed into symmetrical, asymmetrical, and aberrantly-shaped cell clusters that eventually formed embryos at different frequencies. This indicates that organized growth polarity do not always exist in somatic embryogenesis.

Gymnosperms

Embryo development in gymnosperms occurs in three phases. Proembryogeny includes all stages prior to suspensor elongation. Early embryogeny includes all stages after suspensor elongation but before root meristem development. Late embryogeny includes development of root and shoot meristems. Time-lapse tracking in Norway Spruce *Picea abies* revealed that neither single cytoplasmic-rich cells nor vacuolated cells developed into embryos. Proembryogenic masses (PEMs), an intermediate between unorganized cells and an embryo composed of cytoplasmic-rich cells next to a vacuolated cell, are stimulated with auxin and cytokinin. Gradual removal of auxin and cytokinin and introduction of abscisic acid (ABA) will allow an embryo to form. Using somatic embryogenesis has been considered for mass production of vegetatively propagated pine clones and cryopreservation of germplasm. However, the use of this technology for reforestation and breeding of pine trees is in its infancy.

In Vitro Somatic Embryogenesis

In vitro somatic embryogenesis (SE) was first demonstrated in 1958 by Reinert and Steward. There are two ways by which SE could be obtained - i) Indirect SE, where first the callusing is induced from the explant by rapid cell division and later the callus give rise to SE, and ii) Direct SE, where the somatic embryos are developed directly from the explant without an intermediate callus phase.

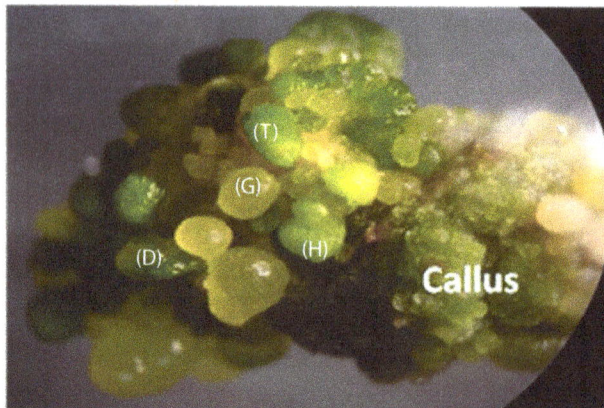

Somatic embryogenesis via callusing showing the development of globular(G), heart (H), torpedo (T) and dicot embryos (D) (arrow marked).

Direct somatic embryogenesis from cotyledon explant showing embryos at various stages of development

In either of the cases, the somatic embryos resemble the zygotic embryos. In dicotyledonous plants, the somatic embryos passes through the globular, heart, torpedo and cotyledonary stages, as happens in zygotic embryos. The embryos germinate and develop into complete plantlets. The only major difference between somatic and zygotic embryogenesis is that somatic embryos do not pass through the desiccation and dormancy phases as happens in zygotic embryos, but rather continue to participate in the germination process.

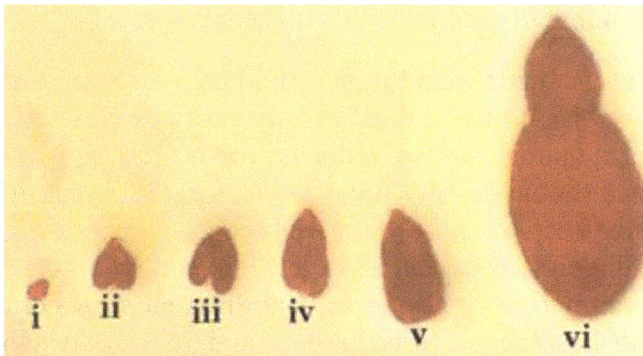

Different stages of development of zygotic embryos: (i) globular, (ii) early heart shape, (iii) late heart shape, (iv) torpedo shape, (v) early dicot, and (vi) fully developed dicot embryo

Whether originating directly or indirectly via callusing, somatic embryos arise from single special cells located either within clusters of meristematic cells in callus mass or in the explant tissue. Somatic embryogenesis is regarded as a three step process:

i. Induction of embryo

ii. Embryo development

iii. Embryo maturation

Organogenesis Versus Embryogenesis

In tissue cultures, plant regeneration via somatic embryogenesis may offer many advantages over organogenesis, such as:

i.   Embryo is a bipolar structure rather than a monopolar one.

ii. The embryo arises from a single cell and has no vascular connection with maternal callus tissue or the cultured explant. On the other hand during organogenesis shoots or roots develop from a group of cells resulting into chimera formation which later establish a strong connection with the maternal tissue.

iii. Further, induction of somatic embryogenesis requires a single hormonal signal to induce a bipolar structure capable of forming a complete plant, while in organogenesis, it requires two different hormonal signals to induce shoot first and then root organ.

Factors Affecting Somatic Embryogenesis

Genotype and Type of Explant

Like organogenesis, SE is also genotype dependent for a given species and significant variations in response between cultivars have been observed in several plants like, wheat, barley, soyabean, rice, alfalfa etc. Genotypic variations could be due to endogenous levels of hormones, therefore, if the species has not shown SE previously, then it is required to test number of different cultivars of that species.

The next problem comes up is what tissue should be used as an explant in a particular species to induce SE? It can be decided very easily by closely examining of what explants were used in related species, genus or family. The explant selection is much more important than the media selection for SE process. Immature zygotic embryos have proved to be the best explant to raise embryogenic cultures as somatic embryos will form more readily from cells that are already in embryonic state. In *Azadirachta indica* (neem), the immature zygotic embryo at different stages of development, viz. globular, early to late heart shape, torpedo shape and early dicotyledonous stage, when cultured showed varied potential for SE. The globular embryos did not show any response. The older embryos germinated, formed calli or differentiated three types of organized structures, viz. shoots, somatic embryos and neomorphs (abnormal or embryo-like structures with varied morphology). Often the same explant differentiated more than one kind of regenerants. The most responsive stage of embryos was early dicotyledonous, followed by torpedo shaped embryos.

Growth Regulators

Auxin : Auxin plays a major role in the development of somatic embryos. All the well-studied somatic embryogenic systems, such as carrot, coffee and most of the cereals require a synthetic auxin for the induction of SE followed by transfer to an auxin-free medium for embryo differentiation. The synthetic auxin 2,4-D is the most commonly used auxin for the induction of SE. Besides, other auxins, NAA, IBA, picloram (4-Amino-3,5,6-trichloro-2-pyridinecarboxylic acid) and IAA, have also been used. A naturally occurring auxin IAA is a weak auxin and more readily broken down compare to 2,4-D and NAA. The auxins, particularly 2,4-D, in the concentration range of 0.5 – 1.0 mgl⁻¹ (proliferation or induction medium), stimulates the formation of localized group of meristematic cells in the callus called 'proembryogenic masses' (PEMs), which are cell clusters within cell population competent to form somatic embryos.

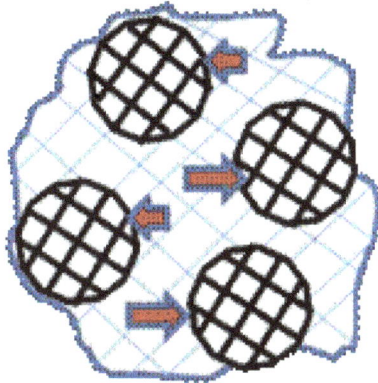

Embryogenic callus with PEMs (indicated by arrows) in the induction medium

In repeated subcultures on the proliferation medium, the embryogenic cells continue to multiply without the appearance of embryos. However, if the PEMs are transferred to a medium with a very low level of auxin (0.01-0.1 mgl⁻¹) or no auxin in the medium (embryo development medium ; ED medium), they develop into embryos. The presence of an auxin in the proliferation medium seems essential for the tissue to develop embryos in the ED medium. The tissues maintained continuously in auxin-free medium would not form embryos. Therefore, the proliferation medium is called the 'induction medium' for SE and each PEMs as an unorganized embryo.

Cytokinin : There are reports of somatic embryo induction and development in cytokinin containing medium, but these reports are very few compared to those reporting induction by auxin alone or auxin plus cytokinin. Cytokinin, in general, induced SE directly without the callusing of explant. In most cases, TDZ is used as cytokinin, a herbicide, which mimics both auxin and cytokinin effects on growth and differentiation. The other cytokinins are also used when zygotic embryos are used as the explant source. The most commonly used cytokinins are BAP and Zeatin.

In *Azadirachta indica*, somatic embryo differentiation was influenced by the culture medium as well as the stage of embryo at culture. Maximum somatic embryogenesis occurred directly from the explant on BAP containing medium when early dicotyledonous stage of embryos were cultured. Medium with 2,4-D induced only neomorph differentiation directly from the explant. While torpedo shaped embryos showed both neomorph formation as well as somatic embryogenesis on BAP containing medium.

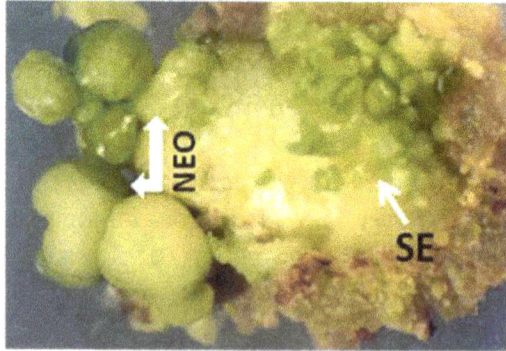

An explant showing differentiation of neomorphs (NEO) and somatic embryos
(SE) on the same explant

Neomorphs were suppressed embryos with green, smooth, shiny surface and solid interior. Although they were epidermal in origin like somatic embryos with heart shape notch but showed monopolar germination and no clear cut radicular region.

A. An explant showing direct differentiation of neomorphs. Some of these structures also show cotyledon-like flaps. The portion of the explant in contact with the medium has proliferated into a brownish green callus

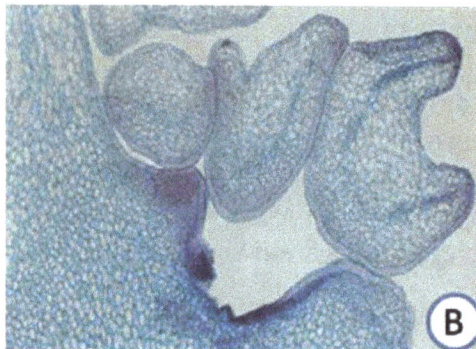

B. A histological section of A, showing epidermal origin of a neomorph of various shapes. It has a well differentiated epidermis and compactly arranged internal cells. These structures are loosely attached to the explant and show provascular strands.

Nitrogen Source

The most important nutrient of the culture medium is nitrogen which affects SE significantly. The form of nitrogen have a strong influence on the induction of SE. Often the presence of ammonium or some other source of reduced nitrogen is required, such as glycine, glutamate or casein hydrolysate. The ratio of ammonium to nitrate has also been shown to affect SE. In few cases, the calli initiated on a medium with KNO_3 as the sole source of nitrogen failed to form embryos upon removal of auxin. However, the addition of a small amount (5mM) of nitrogen in the form of NH_4Cl in the presence of 55mM KNO_3 allowed embryo development.

Embryo Maturation and Germination

Germination of somatic embryos can occur only when it is mature enough to have functional shoot and root apices capable of meristematic growth. Somatic embryos show poor germinable quality with respect to their convertibility into complete plantlets. This is because these embryos do not go through 'embryo maturation' phase which is the characteristic of seed or zygotic embryos. During this phase, accumulation of embryo-specific reserve food materials and proteins imparts desiccation tolerance to seed embryos and thereby promote their normal development for germination. Abscisic acid (ABA), which prevents precocious germination and promotes normal development of embryos by suppression of secondary embryogenesis and pluricotyledony, is reported to promote embryo maturation in several species. A number of other factors, such as temperature, shock, osmotic stress, nutrient deprivation and high density inoculums, can substitute for ABA, presumably by inducing the embryos to synthesize the hormone. ABA is known to trigger the expression of genes which normally express during the drying down phase of seeds. Probably the products of these genes impart desiccation tolerance to the embryos.

Secondary Somatic Embryogenesis

Secondary SE is a process in which new somatic embryos are proliferated from originally formed primary somatic embryos. Secondary SE is reported to have some advantage over primary somatic embryogenesis, such as high multiplication rate, long term repeatability and independency of an explant source. By repeated secondary SE selected embryogenic lines can be maintained for long period, in large quantities until the lines have been tested in field conditions particularly in perennial plants. Secondary SE also overcomes post fertilization barriers of the embryo, immature embryos of interspecific plants from incompatible crosses may be rescued by culturing them for secondary SE. It can also be used for the production of somatic embryos of species in which the embryos are the reservoir of important secondary metabolites. In *Azadirachta indica*, for secondary SE, primary embryos were used as the explant and when cultured on medium with TDZ and GA_3, secondary embryos were differentiated directly from hypocotyls region without any intervening callus. Whereas, a combination of BAP and IAA resulted into secondary SE preceded by callusing of the primary somatic embryo.

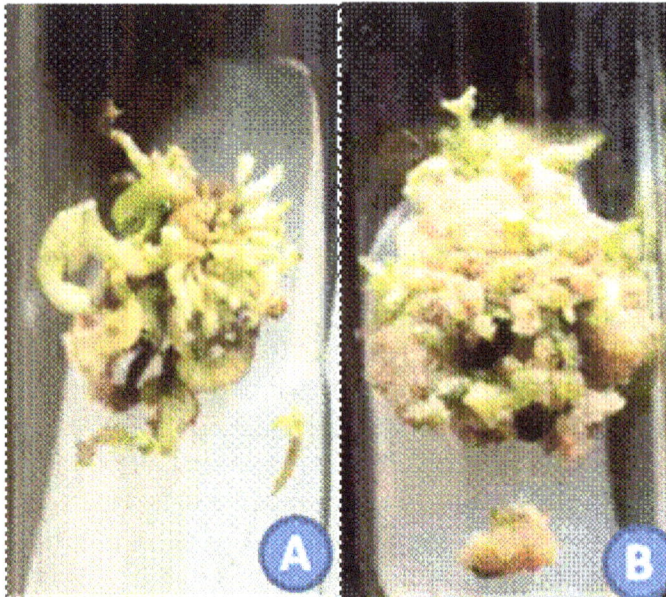

Primary somatic embryo showing: **A.** direct secondary somatic embryogenesis
B. indirect secondary somatic embryogenesis

Synchronization of Embryo Development

Generally, the differentiation of somatic embryos in semi-solid medium or liquid medium is highly asynchronous which adversely affect the germination rate. Synchronization of embryo development is very important for artificial seed technology. Of the several approached tried to achieve this, the most effective method are the physical separation of embryogenic stages and use of growth regulators to physiologically synchronize the development. The other alternative methods are the fractionation of embryos of different stages by filtration of suspension through meshes of different sizes or by gradient centrifugation. Besides, the most effective method to achieve synchronous development of somatic embryos is the use of substances that would induce reversible cessation of embryo development at a particular stage. ABA at low concentration is the most satisfactory substance for the purpose. For example, in carrot it inhibits the growth of roots and enhances suspension with torpedo shaped embryos.

Production of Synthetic Seeds or Artificial Seed

Although it is possible to use naked embryos for large scale planting, it would be beneficial to convert them into 'synthetic seeds' or 'synseeds' for large scale clonal propagation at commercial level. This can be achieved by encapsulating the viable somatic embryos in a protective covering. The coating material should have several qualities:

i.   It must be non-damaging to the embryos.

ii. The coating should be mild enough to protect the embryos and allow germination but it must be sufficiently durable for rough handling during manufacture, storage, transportation and planting.

iii. The coat must contain nutrients, growth regulators and other components necessary for germination.

iv. The quality of somatic embryo should be good enough, they all are of uniform stage with reversible arrested growth and with high rate of conversion to plantlets.

Two types of synthetic seeds are produced:

I.   Desiccated synthetic seeds II. Hydrated synthetic seeds

I.   Desiccated synthetic seeds : It involves encapsulation of somatic embryos followed by their desiccation and can be prepared by following methodology:

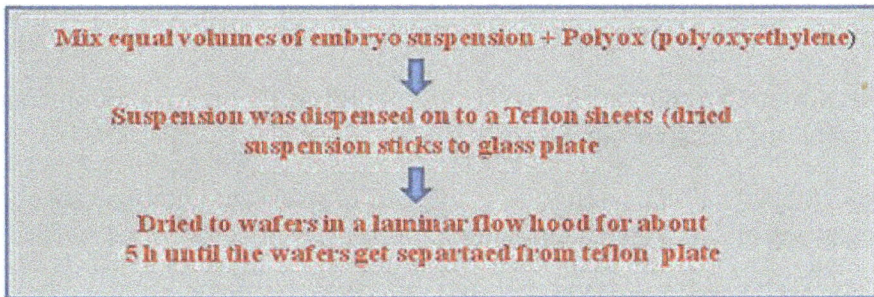

> Mix equal volumes of embryo suspension + Polyox (polyoxyethylene)
>
> ⬇
>
> Suspension was dispensed on to a Teflon sheets (dried suspension sticks to glass plate
>
> ⬇
>
> Dried to wafers in a laminar flow hood for about 5 h until the wafers get separtaed from teflon plate

The polyox is readily soluble in water and dries to thin film. It does not support the growth of microorganism and is non toxic to the embryos. Embryo survival and conversion of seeds are determined by redissolving the wafers in embryogenic medium and culturing the rehydrated embryos.

II. Hydrated synthetic seed: Several methods have been examined to produce hydrated artificial seeds of which Ca-alginate encapsulation has been the most widely used. It can be prepared by following steps:

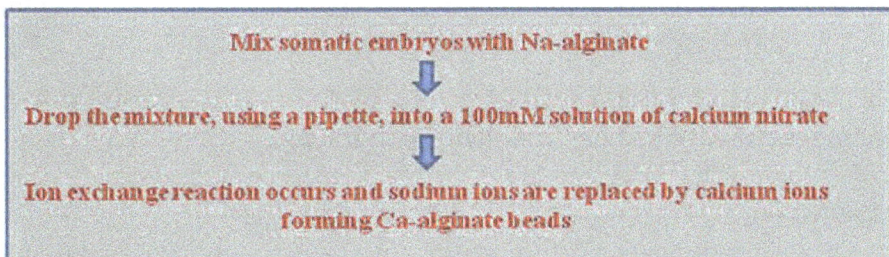

> Mix somatic embryos with Na-alginate
>
> ⬇
>
> Drop the mixture, using a pipette, into a 100mM solution of calcium nitrate
>
> ⬇
>
> Ion exchange reaction occurs and sodium ions are replaced by calcium ions forming Ca-alginate beads

Applications of Somatic Embryogenesis

Following features of somatic embryos prompted many scientists to achieve regeneration via somatic embryogenesis using various explants, most popular ones being zygotic embryos, or excised cotyledons or hypocotyls:

i. Somatic embryogenesis offers immense potential to speed up the clonal propagation of plants being bipolar in nature.

ii. Being single cell in origin, there is a possibility to automate large scale production of embryos in bioreactors and their field planting as synthetic seeds.

iii. The bipolar nature of embryos allows their direct development into complete plantlet without the need of a rooting stage as required for plant regeneration via organogenesis.

iv. Epidermal single cell origins of embryos favor the use of this process for plant transformation.

v. It can also be used for the production of metabolites in species where embryos are the reservoir of important biochemical compounds.

vi. The production of artificial seeds using somatic embryos is an obvious choice for efficient transport and storage.

vii. The embryo culture technique is applied to overcome embryo abortion, seed dormancy and self-sterility in plants.

Limitations of Somatic Embryogenesis

i. Complete conversion into plantlets or poor germination of embryos is a major limitation of somatic embryogenesis in many plants. Therefore, the process of germination needs to be studied in detail for successful plantlet conversion.

ii. Compared to other plant species active research on somatic embryogenesis involving forest trees has been very slow.

iii. The paucity of knowledge controlling somatic embryogenesis, the synchrony of somatic embryo development and low frequency of true to type embryonic efficiency are responsible for its reduced commercial application

iv. To obtain a complete conversion into plantlets it is necessary to provide optimum nutritive and environmental conditions.

Androgenesis

In androgenesis, the male gametophyte (microspore or immature pollen) produces haploid plants. The basic principle is to stop the development of pollen into a gamete (sex cell) and force it to develop into a haploid plant or sporophyte. The remarkable discovery that haploid embryos and plants can be produced by *in vitro* culture of anthers of *Datura* (Guha and Maheshwari 1964,1966) brought renewed interest to haploidy. This method of androgenic haploid production was quickly attempted in many species to hasten the breeding programme in several economically important plants. Haploid production through anther/microspore culture scores higher over other methods due to the fact that anthers harbour large numbers of haploid microspores per anther and is a potentially efficient means to generate homozygous true-breeding progeny lines in plant breeding programs.

Methodologies

In androgenesis, immature pollen grains are induced to follow the sporophytic mode of development by the application of various physical and chemical stimuli. There are two methods for *in vitro* production of androgenic haploids –

(A) Anther culture, and (B) Isolated pollen (microspore) culture.

A. Anther culture

(a) Select the flower buds from an elite plant and determine the stage of microspores by acetocarmine squashes of anthers or by staining them with fluorescent dye DAPI (4,6-diamidino-2-phenylindole).

(b) Surface sterilize the selected size of buds (ca 2mm size flower bud) to initiate in vitro anther cultures.

(c) Dissect the buds under a stereo-microscope, using pre-sterilized Petriplates, forceps and fine needles. Discard the damaged anthers, if any, and remove the filament gently.

(d) Inoculate the anthers, bearing early-to-late uninucleate stage of microspores, in a nutrient medium and maintain the cultures in defined conditions.

(e) As the anthers proliferate, they produce embryos/callus.

(f) The callus/embryos formed can be transferred to a suitable medium to finally produce haploid plants and then diploidize them by colchicine to produce homozygous diploids.

B. Microspore (Pollen) culture

Haploid plants can also be produced from isolated immature pollens or microspores (male gametophytic cells):

(a) Select the flower buds from an elite plant and determine the stage of microspores by acetocarmine squashes of anthers or by staining them with fluorescent dye DAPI (4,6-diamidino-2-phenylindole).

(b) Surface sterilize the selected size of buds to initiate *in vitro* anther cultures.

(c) Extract the microspores by pressing and squeezing the buds with a glass rod against the sides of a beaker.

(d) Filter the pollen suspension to remove anther tissue debris.

(e) Wash and collect the viable and large pollen (smaller pollen do not regenerate) by filtration.

(f) Culture these microspores on a solid or liquid medium.

(g) As the microspores undergo multiple divisions, they produce multicellular and multinuclear structure.

(h) The callus/ embryos formed can be transferred to a suitable medium to finally produce a haploid plants and then diploid plants by colchicine treatment.

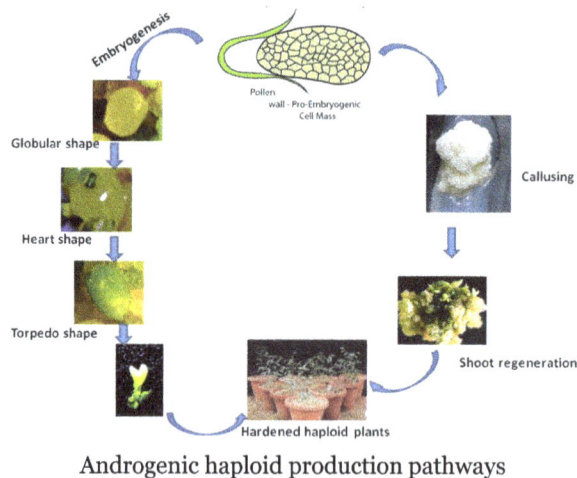

Androgenic haploid production pathways

Comparison between Anther and Pollen Culture

Anther culture is an easy, quick and practicable approach. Anther walls act as conditioning factors and promote culture growth. Thus, anther cultures are reasonably efficient for haploid production. The major limitation is that the plants not only originate from pollen but also from other somatic parts of the anther. This results in the development of plants at different ploidy levels viz., haploids, diploids, aneuploids, as a result of which the final tissue, derived, may not be of purely gametophytic origin. Moreover, the plants arising from an anther would constitute a heterogenous population. It has been observed in some species that anther cultures show asynchronous pollen development, the older grains may suppress the androgenic potential of younger grains by releasing toxic substances . The disadvantages associated with anther culture can be overcome by pollen culture as it offers the following advantages:

- Undesirable effects of anther wall and associated tissues can be avoided.

- Androgenesis, starting from a single cell, can be better regulated.

- Isolated microspores (pollen) are ideal for various genetic manipulations like transformation, mutagenesis etc.

- The yield of haploid plants is relatively higher.

Pathways of Development

The early divisions in responding pollen grains may occur in any one of the following four pathways:

i)   Pathway I - The uninucleate pollen grain may divide symmetrically to yield two equal daughter cells both of which undergo further divisions e.g. *Datura innoxia* .

ii) Pathway II - In some other cases e.g. *N.tabacum*, barley, wheat etc., the uninucleate pollen divides unequally. The generative cell degenerates, callus/embryo originates due to successive divisions of the vegetative cell.

iii) Pathway III - But in few species, the pollen embryos originate from the generative cell alone; the vegetative cell either does not divide or divides only to a limited extent forming a suspensor like structure.

iv) Pathway IV - Finally in few other species e.g. *Datura innoxia*, the uninucleate pollen grains divide unequally, producing generative and vegetative cells, but both these cells divide repeatedly to contribute to the developing embryo/callus.

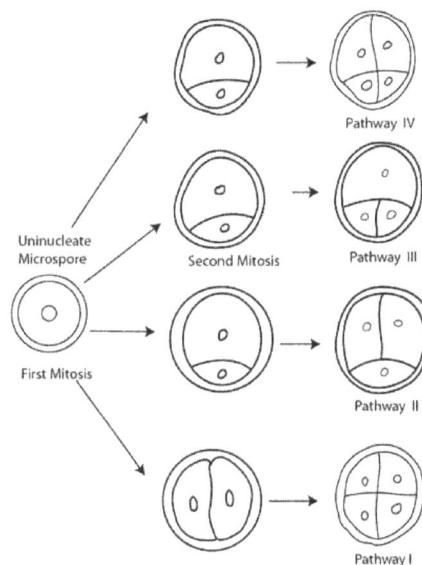

Pathways of development of microspores

Gynogenesis

Gynogenic development of plants from unfertilized cells of female gametophyte (embryo-sac) in ovary/ovule/young flower cultures is one of the available alternatives for haploid production. It was first reported in barley San Noeum (1976). This method of haploid production is more tedious than androgenesis. The reasons for this being the indefinite numbers of microspores (male gametes) within the anther wall for androgenesis as against single egg cell (female gamete) per flower for gynogenic haploid production, which too, is deep seated within the embryo-sac (female gametophyte), thus making the entire process very cumbersome. The technique is very useful where anther culture has been unsuccessful, plants are male sterile or androgenesis is confronted with the problem of albino or non-haploid formation. The following techniques are generally used for production of haploids via *in vitro* gynogenesis either through direct embryogenesis or via callusing.

In Situ Parthenogenesis Induced by Irradiated Pollen Followed by in Vitro Embryo Culture

Parthenogenesis induced *in vivo* by irradiated pollen, followed by *in vitro* culture of embryos can be an alternative method of obtaining haploids in fruit crops. Gynogenesis by *in situ* pollination with irradiated pollen has been successfully used for *Malus domestica* (L.) Borkh, *Pyrus communis* L., *Actinidia deliciosa* (A. Chev). This method is based on *in vitro* culture of immature seeds or embryos, obtained as a result of pollination by irradiated pollen with gamma rays from cobalt 60. The method is useful in those species in which *in vitro* anther culture has not been successful. Irradiation does not hinder pollen germination but prevents pollen fertilization and, thereby, stimulating the development of haploid embryos from ovules. The success of this technique is dependent on the choice of radiation dose, the developmental stage of the embryos at the time of culture, culture conditions and media requirements.

Ovary Slice Culture

Ovary slice culture technique involves culture of transverse sections of unpollinated ovaries on culture media. The following protocol was used to induce *in vitro* gynogenesis in Tea (Hazarika and Chaturvedi 2012):

a. For Ovary slice culture in Tea, unopened and unpollinated mature flower-buds (6-10 mm) size were collected early in the morning. Some of the buds were fixed in FAA (5:5:90 v/v/v Formaldehyde: Acetic acid: 70% Ethanol), for 48 h, and then stored in 70% alcohol. Later on, the appropriate developmental stage of the embryo sac was determined by histological analysis.

b. The flower buds were surface sterilized with 0.1% $HgCl_2$ for 7 minutes, followed by rinsing with sterile distilled water at least thrice.

c. Carefully dissected transverse sections of ovaries were cultured on Murashige and Skoog's media supplemented with varying concentrations of Auxins and Cytokinins.

d. Six ovary slices containing unpollinated ovules were cultured in 60 mm X 15 mm pre sterilized disposable Petriplates containing 10 ml MS medium.

e. The sealed Petriplates were subjected to various regimes of temperature and light treatments.

Ovule Culture

The unfertilized ovary is surface sterilized and the ovules were taken and placed into culture. Excision of ovule, followed by culture on specific media may be either extremely easy to accomplish, as in case of large-seeded species in which only a single ovule is present, or time-consuming and intricate, in small-seeded polyovulate species. Two types of ovule support systems have been developed. The filter paper support system involves culturing of the ovules on top of filter paper placed over liquid medium, whereas the vermiculite support technique demands placing the ovules on a sterile vermiculite/liquid media mixture (vermiculite support) with the micropylar side down. Unpollinated ovule culture has been used for haploid production in sugar beets and onions.

Factors Affecting Haploid Androgenesis

There are numerous endogeneous and exogeneous factors that affect *in vitro* haploid production. These factors can be genetic, physiological, physical and chemical may also interact amongst each other to divert the microspores/egg cell to enter into a new developmental pathway. Some of the crucial factors affecting haploid production in plants have been discussed below:

Genotype of the Donor Plant

The genotype of the donor plants, i.e. genetic factor, has a great influence on the anther, ovary and ovule culture response. In earlier studies, significant difference in callus formation using varieties or crosses were observed. In some species only a few genotypes have responded of many tested. In fact genetic factors contribute in a major way to the differences in the number of haploid plants produced (Custódio et al, 2005).

Physiological Status of the Donor Plant

The physiological conditions of the donor plant, *i.e* the environmental conditions and age of the donor plant, directly affects both *in vitro* androgenesis and *in vitro* gynogenesis in almost all plant species. A correlation between plant age and anther response has also been demonstrated. Similar is the case with ovary culture. The frequency of androgenesis is usually higher in anthers harvested at the beginning of the flowering period and showed a gradual decline in relation to plant age (Bhojwani and Razdan 1996). Varying temperature and light conditions during the growth of donor plants also affect anther response. In anther culture of grape, the induction frequency of embryoids derived from spring flowers was higher than that derived from summer flowers (Zhou and Li 1981). The microscopical observations showed that some varieties of rubber tree often have a lot of degenerated and sterile microspores in their anthers in early spring or hot summer due to the influence of unfavourable climatic conditions (Chen et al, 1982). As a result no pollen embryoids were obtained from such anthers but only the somatic calli.

Stage of Explants Material at the Time of Inoculation

Stage of Microspores

The stage of microspores at the time of inoculation is one of the most critical factors in the induction of androgenesis. Detailed cytological studies conducted on poplars, rubber (Chen 1986) and apple (Zhang et al, 1990) have shown that androgenic callus and embryos were mainly induced through a deviation of the first pollen mitosis to produce two undifferentiated nuclei. Besides affecting the overall response, the microspore stage at culture also has a direct bearing on the nature of plants produced in anther culture (Sunderland and Dunwell 1977). About 80% of the embryos obtained from binucleate microspores of *Datura innoxia*, a highly androgenic species, were non-haploids (Sunderland et al, 1974). In a vast majority of species where success has been achieved, anthers were cultured when microspores were at the uninucleate stage of microsporogenesis (Chaturvedi et al, 2003; Pedroso and Pais 1994; Sopory and Munshi 1996).

Stage of the Embryo Sac

It has been reported that the effect of ovule development on gynogenesis is profound as it harbours the embryo sac comprising of the egg cell. The stage of embryo sac is an important determining factor for *in vitro* gynogenesis in various plant species. However, it is difficult to know the stage of embryo sac at the time of inoculation. Several authors prefer to describe the inoculation stage according to the developmental stage of the flower bud or stage of pollen development. However, this could not be possible in several species, where male and female gametophytes do not mature simultaneously, a phenomenon known as protandry, the maturation of anthers before carpels (e.g., onion, leek, sunflower, sugar beet and carrot) and the opposite protogyny (e.g., pearl millet). In such cases, the stage of embryo-sac at culture can be determined by histological preparations of ovary/ovules that are at identical stage with that of cultured ovary/ovules.

Although a wide range of embryo sac stages are responsive to gynogenic development, but, in most cases nearly mature embryo-sac stage gave better results. This is quite contrary to anther culture in which mature pollen is non responsive to androgenesis.

In Barley and rice, unfertilized ovary cultures with late staged mature embryo sacs gave good results (San Noeum 1976, 1979; Wang and Kuang 1981) while others reported success with ovary cultures containing uninucleate to mature embryo sacs (Zhou and Yang, 1981b, 1982; Kuo, 1982; Huang et al., 1982).

Anther Wall Factor

One of the important research subject in anther culture of woody plants is to avoid the over proliferation of callus from anther wall tissues and to achieve a high yield of pollen embryoids and pollen calli. In anther culture of most woody plants, both pollen calli (or embryoids) and somatic calli from anther wall tissues grew simultaneously. The development of callus from somatic tissues of anther can be avoided by culture of isolated microspores. However, microspore culture is not very successful in woody plants (Chaturvedi et al, 2003). Pelletier and Ilami (1972) had introduced the concept of "Wall Factor", according to which the somatic tissues of anther play an important role in the induction of sporophytic divisions in pollen. In anther culture of rubber, 47% of the microspores in close contact with the surrounding somatic cells could develop into multicellular masses as compared to only 5% of microspores away from the wall (Chen 1986). Anther wall callusing was regarded as a pre-requisite for the formation of androgenic haploids (Chaturvedi et al, 2003; Chen 1986; Chen et al, 1982).

Chemical Factors

The constituents of the basal medium and combination/s of growth regulators serve as important factors in eliciting successful androgenesis or gynogenesis. However, it is difficult to suggest one culture medium or one growth regulator for all the systems. The requirement of growth regulators and culture medium in terms of kind and concentration may vary with the system. Generally, there is an agreement that the source and amount of total nitrogen as well as a combination of a cytokinin and an auxin is necessary for pollen embryogenesis and pollen callusing in several woody plants (Chaturvedi et al, 2003; Chen 1986; Nair et al, 1983). Sucrose has generally been used as the major carbohydrate source in the culture medium. The effect of sucrose on anther culture has been investigated in a number of species. Generally, sucrose is supplied at 2-3% concentration. However, increase in its concentration can lead to beneficial morphogenic potential by suppressing the proliferation of somatic tissues.

Effect of Female Flower Position

Position of female flowers on the plant stem affected induction of embryos from ovule cultures of *Cucurbita pepo* L (Shalaby, 2007). One of the possible explanations for enhancing responses of tissue culture could be attributed to indigenous hormonal level (Johansson, 1986).

Diploidization of Haploids

The haploids may grow normally up to flowering stage but in the absence of homologous chromosomes the meiosis is abnormal and consequently, viable gametes are not formed. To obtain fertile homozygous diploids, the chromosome complement of the haploids must be duplicated. For long colchicine has been used for the purpose and is more effective.

Another method to diploidise the haploids is by utilizing the tendency of *in vitro* growing callus cells to undergo endomitosis. Segments from vegetative parts of haploid plants are grown on a suitable culture medium and made to proliferate into calli. After some time many of the callus cells become diploid due to endomitosis. By transferring such calli to an appropriate medium fertile diploids shoot can be obtained.

Applications of Haploid Production

Development of Pure Homozygous Lines

Homozygous, true breeding cultivars are highly important for screening of high yielding lines and to produce hybrid vigour as a method of crop improvement. Obtaining homozygous diploid plants by conventional methods is difficult in perennials. From several decades to over a hundred years are required to obtain a pure line by means of successive inbreeding throughout many generations. The seed set by inbreeding in many trees is very low, usually only a few of ten thousandth or sometimes no seed can be obtained at all; therefore, it is impractical to obtain pure lines by inbreeding (Chen, 1986). Moreover, conventional method of haploid production by inbreeding is impossible if the plant is strictly cross-pollinating in nature. On the other hand, homozygous diploid plants can be achieved in a single generation by diploidization of *in vitro* raised haploids by colchicine treatment.

Genetic Studies

Because of the lack of accurate materials in research work, the progress in the study of genetics in trees is much slower than that in annual herbaceous plants. The genetics of a lot of important traits in economically important plant species has not been clearly demonstrated as yet. As a result, it is still unknown whether the desirable characters of the parents will appear in their progenies. Only when crossing between different homozygous diploid plants is carried out, we can gain a clear idea of dominance of genes controlling various characters and that these characters are either monogenic or polygenic (Chen 1986). Furthermore, if we can use the haploid plants as samples of gametes, then we can obtain directly the recombination value between genes. Moreover, we can also use the haploid plants to study chromosome homology within genome or between genomes.

Gametoclonal Variation

The "gametoclonal variation" arises among plants regenerated from cultured gametic cells consisting of differences in morphological and biochemical characteristics as well as in chromosome number and structures that are observed. Besides yielding haploids, *in vitro* androgenesis helps in the screening of gametophytic variation at plant level. Pollen grains within an anther form a highly heterogeneous population because they are the product of meiosis which involves recombination and segregation. Therefore, each pollen plant is genetically different from the other. The gametoclonal variants being hemizygous in nature expresses also the recessive characters in the R_0 plants (Bhojwani and Razdan, 1996). Different sources of variation can explain gametoclonal variation such as new genetic variation induced by cell culture procedures, from segregation and independant assortment, chromosome doubling procedures etc (Morrison and Evans, 1987; Huang, 1996).

Induction of Mutations

In general, majority of induced mutations are recessive and, therefore, are not expressed in diploid cells due to the presence of dominant allele. Since, haploid plants have only one set of chromosomes, their dominant and recessive characters can be seen simultaneously on separate plants. It is extremely advantageous to provide a convenient system for the induction of mutations and selection of mutants with desirable traits in the absence of their dominant counterparts (Bhojwani and Razdan, 1996)

Obtaining New Genotypes with Alien Chromosomes

The technique of interspecific and intergeneric hybridization can be combined with anther culture techniques (Thomas et al., 2003) for obtaining new genotypes with alien chromosomes. Thus, new genotypes with various reconstructed chromosome complements can be obtained after their successful chromosome doubling.

Genetic Manipulation

As microspore culture is a single cell system, it makes selection at the single cell level possible and, furthermore, offers new prospects for genetic manipulation like mutagenesis and transformation. Direct gene transfer by microinjection offers the possibility of transgenic plant formation by using isolated pollen culture having high regeneration efficiency (Kasha and Maluszynski, 2003). Moreover, if transgenes can be incorporated into the haploid microspore genome, prior to DNA synthesis and chromosome doubling, the doubled haploids may also be homozygous for the transgenes. Thus, isolated microspores not only provide a good target for bombardment but, also are readily amenable to transgene *in vitro* selection. Jahne et al. (1994) were the first to achieve plants homozygous for the transgenes using biolistic bombardment of barley microspores.

Genomics

Doubled haploids play a vital role in genomics, especially, in the integration of genetic and physical maps, thereby, providing precision in targeting candidate genes (Kunzel et al., 2000; Wang et al., 2001). Doubled haploids combined with marker assisted selection provides a short cut in backcross conversion, a plant breeding method for improving an elite line defective in a particular trait (Toojinda et al., 1998). Expressed sequence tags may help in identification of genes that determine any trait of interest.

Triploid Production by Endosperm Culture

Introduction

Endosperm is a unique tissue in its origin, development and ploidy level. It is a product of double fertilization but unlike the embryo it is triploid and develops into a formless tissue (Bhojwani and Bhatnagar 1999). It is, therefore, an interesting tissue for morphogenesis. Any abnormality in the development of endosperm may cause the abortion of embryo resulting in sterile seeds (Johnston et al. 1980). The endosperm may be totally consumed by developing embryo leading to the formation of exalbuminous (non-endospermous) seeds or when it persists, the seed is called albuminous (endospermous). In albuminous seeds, it is used as a food source which may contain proteins, starch or fats and the embryo can utilize this food during seed germination.

Cellular totipotency of endosperm cells was first demonstrated by Johri and Bhojwani in 1965. To date, differentiation of shoots/embryos/plantlets from endosperm tissue has been reported for more than 64 species belonging to 24 families. In many of these reports the regenerants were shown to be triploid. A key factor in the induction of cell divisions in mature endosperm cultures is the association of embryo. The embryo factor is required only to trigger cell divisions; further growth occurs independent of the embryo. Triploid plants are usually seed-sterile. However, there are many examples where seedlessness caused by triploidy is of no serious concern or, at times, even advantageous. Some of the crops where triploids are already in commercial use include several varieties of apple, banana, mulberry, sugar beet and watermelon. Natural triploids of tomato produced larger and tastier fruits than their diploid counterparts (Kagan-Zur et al. 1990).

Traditionally, triploids are produced by crossing induced superior tetraploids and diploids. This approach is not only tedious and lengthy (especially for tree species) but in many cases it may not be possible due to high sterility of autotetraploids. The first step in the process is to produce tetraploids by colchicine treatment of germinating seeds, seedlings or vegetative buds. In most of these cases the rate of induction of tetraploids had been low (7-22%). Moreover, the treatment is lengthy and laborious. Once tetraploids have been produced, their cross with the diploid parent may not be successful in majority of the cases. In successful crosses the seed-set, seed germination and survival rate of the seedlings is low. Moreover, all sexually produced triploids do not behave uniformly, which may be due to segregation both at tetraploid level and subsequent population of crosses with putative diploid. In contrast, *in vitro* regeneration of plants from endosperm, a naturally occurring triploid tissue, offers a direct, single step approach to triploid production. The selected triploids, expected to be sexually sterile, can be bulked up by micropropagation.

Factors Controlling Callus Proliferation and Plant Regeneration

Endosperm at Culture

Usually culture of endosperm needs the selection of seeds at proper stage of development. This is usually calculated as days after pollination (DAP) and it varies from plant to plant as 9-10 days after pollination (DAP) in *Lolium perenne* (Norstog 1956), 8-11 DAP in *Zea mays* (Tamaoki and Ullstrup 1958), 8 DAP in *Triticum aestivum* and *Hordeum vulgare* (Sehgal 1974) and 4-7 DAP in *Oryza sativa* (Nakano et al. 1975). Usually free nuclear endosperm did not produce any callus and the intensity of response depends on the level of organization of endosperm cells.

Plant Growth Regulators and Other Supplements

Selection of a suitable basal medium and the addition of proper growth regulators and other adjuvants are the decisive factors that determine the success of triploid plant development. The culture of immature endosperm requires yeast extract (YE), casein hydrolysate (CH), coconut milk (CM), corn extract (CE), potato extract (PE), grape juice (GJ), cow's milk (CWM) or tomato juice (TJ) despite a suitable medium and growth regulators. Murashige and Skoog (1962) basal medium was mostly used to initiate and improve the response in *in vitro* endosperm cultures. White (1963) basal medium (WM) was also employed in some cases. La Rue (1949) used various organic supplements like CE, PE, TJ, GJ, YE or CWM to raise endosperm callus cultures. Of these, TJ was found to be superior over other additives due to cytokinin-like activities.

Later it was found that the TJ could be replaced by YE. YE induced callus proliferation was reported in *Zea mays*, *Croton*, *Jatropha panduraefolia*, *Lolium*, *Ricinus communis*, *Oryza sativa*, *Coffea arabica*, and *Juglans regia*. Other additives like CH in *Zea mays*, *Exocarpus cupressiformis*, *Dendrophthoe falcata*, *Nuytsia floribunda*, *Putrangiva roxburghii*, *Hordeum vulgare*, *Achras sapota*, *Citrus grandis*, *Prunus persica*, *Actinidia chinensis*, *Actinidia urguta*, *A. deliciosa* and CM in *Acacia nilotica*, and *Codiaeum variegatum* were also employed by different workers.

Most of the immature endosperm needs the presence of one or more growth regulators for plant regeneration except in few , where MS basal medium is sufficient for endosperm embryogenesis. In majority of reports, an auxin, preferably 2,4-D is necessary for callus induction from immature endosperms. In case of mature endosperm, the optimum callus growth was observed either on a cytokinin or a cytokinin in combination with an auxin and for autotrophic taxa, cytokinin, auxin, CH or YE is necessary. In most of the cases the time required to initiate proliferation varies from 10 days to 20 days, but pre-soaking of endosperms with GA$_3$ have reduced the time period from 10 days to 7 days.

Physical Factors

This includes effect of temperature, light and pH on endosperm proliferation. Straus and La Rue (1954) observed that corn endosperms develop better in dark than light conditions. But in *Ricinus* reverse is the case where a continuous light period was found optimum for endosperm proliferation (Srivastava 1971). In some cases, the endosperms were cultured along with the embryo and kept in the diffuse light with 16 h photoperiod. Light conditions facilitate the early germination of embryo and the embryos can be removed easily due to their characteristic green colour (Thomas et. al. 2000). In coffee, the endosperm callus grows better under 12 h light/dark conditions (Keller et al. 1972). In *Lolium* the light doesn't have any significant role on endosperm proliferation (Norstog 1956).

Not much research has been carried out till date with regard to the effect of temperature and pH on endosperm proliferation. In available literature the optimum temperature for endosperm growth was reported to be 25°C (Johri and Srivastava 1973). The pH varies from 4.0 for *Asimina* to 5.0 for *Ricinus* , 5.6 for *Jatropa* and *Putranjiva* and 6.1 for *Zea mays* . In general, 5.5 to 5.8 pH seem to support the best growth of endosperm tissues in cultures.

The Embryo Factor

There is an absolute necessity of the so called "embryo factor" for the proliferation of endosperm (Bhojwani 1968; Srivastava 1971a, b). Some factors contributed by the germinating embryo is required for the stimulation of mature and dried endosperms of few plant species. In general, it has been found that mature endosperm requires the initial association of embryo to form callus but immature endosperm proliferates independent of the embryo. However, in neem the association of the embryo proved essential to induce callusing of immature endosperm; the best explant was immature seeds (Chaturvedi et al. 2003). Similar observation for mulberry was reported by Thomas et al. (2000). However, the embryo factor can be overcome by the use of GA$_3$ as was observed in *Croton bonplandianum* and *Putranjiva roxburghii* (Srivastava 1973). It is reported that during germination, the embryo releases certain gibberellin like substances, which may promote the endosperm proliferation (Ogawa 1964; Ingale and Hageman 1965). However the mature endosperms of *Achras sapota, Santalum album, Emblica officinalis* and *Juglans regia* (Cheema and Mehra 1982; Tulecke et. al. 1988) could proliferate without the association of embryo or pre-soaking of them in GA$_3$.

Shoot Regeneration

Organogenesis from endosperm tissue was first reported in *Exocarpus cupressiform-*

is (a member of the family Santalaceae) by Johri and Bhojwani (1965). The pathway of plant regeneration includes shoot-bud differentiation or embryogenesis directly from the explants or indirectly from proliferating callus. In almost all the parasitic angiosperms, the endosperm shows a tendency of direct differentiation of organs without prior callusing, whereas in the autotrophic taxa the endosperm usually forms callus tissue followed by the differentiation of shoot buds, roots or embryos. Direct shoot regeneration from the cultured endosperm was observed in a number of semi-parasitic angiosperms including *Exocarpus*, *Taxillus*, *Leptomeria*, *Scurrula* and *Dendrophthoe* .

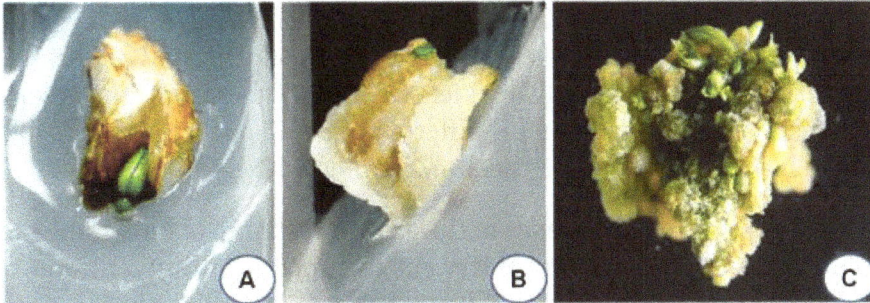

Shoot regeneration from endosperm callus of *Azadirachta indica* : **A.** An immature seed in culture has split open after 2 weeks and releasing the green embryo and callused endosperm ,
B. White fluffy endosperm callus can be seen from the fully opened seed after three weeks. Embryo is lying at one end of the explant, **C.** 6 -week-old subculture of endosperm callus showing the differentiation of distinct shoots and nodules as well.

In *Exocarpus*, an auxin (IAA) along with cytokinin (Kinetin) was required for direct shoot regeneration (Johri and Bhojwani 1965). Addition of zeatin in WM gave rise to green shoots from the intact seed (i.e. endosperm with embryo) culture of *Scurrula pulverulenta* which on subculture gave rise to characteristic haustoria (Bhojwani and Johri 1970). In *Taxillus vestitus*, shoot bud formation occurred on WM supplemented with IAA, Kinetin and CH, after seven weeks. Replacement of IAA with IBA could induce shoot regeneration in 55% cultures and haustoria in 60% cultures (Johri and Nag 1970). Here, the embryo had an adverse effect on bud differentiation from endosperm. Injury to the endosperm was found to be beneficial for shoot induction in some cases; shoot buds first develop along the injured region. The position of the explant on medium plays a significant role in regeneration of shoot in *Taxillus* spp. When half split *T. vestitus* endosperm without embryo was placed on medium with its cut surface in contact with the medium containing Kinetin, 100 % of the cultures produced shoots. In *Leptomeria acida*, IBA proved more efficient than IAA in terms of rapid callus proliferation. However, on IAA supplemented medium the callus gave rise to shoots in 100% cultures (Nag and Johri 1971).

The callus proliferation from endosperm and the subsequent shoot organogenesis was also reported in *Jatropa panduraefolia, Putranjiva roxburghii, Codiaeum variegatum, Malus pumila, Oryza sativa, Annona squamosa, Actinidia chinensis, Mallotus philippensis, Actinidia deliciosa, Morus alba, Azadirachta indica* and *Actinidia deliciosa*. In *Actinidia* species callus initiation occurred on MS medium supplemented with 2,4-D and Kinetin. Transfer of these calli to MS medium containing IAA and 2ip resulted in shoot and root organogenesis. In apple, endosperm proliferated into callus on MS medium supplemented with Kinetin + 2,4-D/BA + NAA and subsequent regeneration occurred on MS medium fortified with BA + CH. In *Annona squamosa* the callusing of endosperm occurred on WM supplemented with two cytokinins (Kinetin and BA), an auxin (NAA) and Gibberellic acid (GA_3). But organogenesis in the callus occurred on Nitsch's medium supplemented with BA and NAA.

In *Mallotus philippensis,* a continuously growing callus was obtained on MS medium supplemented with 2, 4-D + Kinetn. These calli when subcultured on MS + BA + CH gave rise to various morphologically distinct cell lines, of these, only the green compact cell line was responsive for organogenic differentiation. Shoot regeneration occurred in this callus when subcultured on MS medium fortified with BA + NAA.

In Rice, there was a striking difference in the growth response of immature and mature endosperm. Immature endosperm underwent two modes of differentition i.e. direct regeneration from the explant or indirectly via intervening callus phase. In mature endosperm, shoot organogenesis was always preceded by callusing. Callus from mature endosperm was initiated and maintained on MS + 2,4-D; shoot differentiation from callus occurred on MS + IAA + Kinetin. The proliferation of immature endosperm and occasional shoot formation occurred on YE supplemented medium; addition of IAA and Kinetin improved the response further.

Immature endosperms of neem (*Azadirachta indica*) showed best callusing on MS + NAA + BA + CH. When the callus was transferred to a medium containing BA or Kinetin, shoot buds differentiated from all over the callus. Maximum regeneration in terms of number of cultures showing shoot-buds and number of buds per callus cultures occurred in the presence of BA. Thomas et al. observed maximum callusing of Mulberry endosperm on MS + BA + NAA + CH or YE. Shoot buds were emerged when the callus was transferred on a medium containing a cytokinin or a cytokinin and a - naphthaleneacetic acid (NAA). The percent response was highest on BA and NAA containing medium. However, the number of shoots per explants was maximum when TDZ alone was used.

Histology

Histological studies of the proliferating endosperm of *Jatropa, Putranjiva* and *Ricinus*, revealed that the embryo also enlarged and proliferated along with the endosperm but soon showed the sign of degeneration. In such cases the endosperm calli were transferred to a fresh medium to avoid any contamination from degenerated embryonal cells. The 4-week-old callus derived from endosperm cultures, proliferated into parenchymatous cells and 6-week old callus showed tracheidal cells (Srivastava 1971a, b; Srivastava 1973;

Johri and Srivastava 1973). In *Santalum*, endosperm proliferation started after the formation of several meristematic layers below the epidermal region (Rangaswamy and Rao 1963). By carefully applying plant growth regulators the nodular outgrowths can be induced on the surface of the cultured endosperm as in case of *Osyris wightians* (Johri and Bhojwani 1965) and *Putranjiva roxburghii* (Srivastava 1973).

The importance of tracheidal differentiation in the callus of endosperm cultures is that it facilitates organogenic differentiation. In the families like Euphorbiaceae, Loranthaceae and Santalaceae, the endosperm tissues readily form tracheidal elements in cultures (Johri and Srivastava 1973, Johri and Bhojwani 1971). In *Emblica officinale*, tracheidal cells and cambium like cells organized into vascular strands or nodules in the differentiation medium while in the callusing medium tracheidal cells remained scattered. The differentiation of vascular strand in the callus accompanied the shoot bud formation (Sehgal and Khurana 1985).

In *Aleuritus fordii*, callus proliferated from endosperm explant consisted of large, compact and vacuolated cells. Tiny group of cells became distinct from adjoining large and vacuolated cells and became meristematic. These cells remained thin walled with dense cytoplasm and a large clear nucleus. Later the meristemoids developed in to dome shaped shoot apex, which produces leaf primordia (Syed Abbas 1993). In *Mallotus phillippensis* only the compact green callus underwent differentiation. Such callus showed vasculature, developed protuberances and eventually gave rise to shoots buds. Small group of cells with deep-seated distinct meristematic loci were also observed in these calli, which later gave rise to dome-shaped shoot primordia, endogenously (Sehgal and Abbas 1996).

In mulberry (*Morus alba*), histological analysis revealed that the older region of the callus comprised of highly vacuolated cells. Shoot buds differentiated from peripheral nodular structures, which comprised of compactly arranged highly cytoplasmic cells. Often a few layers of degenerating vacuolated cells were seen outside the shoot primordia. It is possible that the shoots originated from inside the nodules and emerged after rupturing the surrounding tissue. The regenerating shoots showed vascular supply continuous with the vasculature of the callus (Thomas 2000).

Both exogenous and endogenous differentiation of shoots was observed in *Azadirachta indica* . The serial section of two-week-old regenerating callus showed that many meristematic pockets developed from inside the callus, which developed into shoot buds after 3 weeks. Histological sections also revealed that the shoot buds emerged from the peripheral tissues of the callus as well (Chaturvedi et al. 2003). In *Actinidia deliciosa* histological analysis of the freshly isolated endosperm revealed small intercellular spaces and cells were filled with storage materials. However, the calli derived from the endosperm were larger, vacuolated and lacked storage materials. In older callus daughter nuclei attached to newly formed cell walls were often observed, suggesting disturbances of cell division. The cells differed in size and shape and contained nuclei with variable numbers of nucleoli (Goralski et al. 2005).

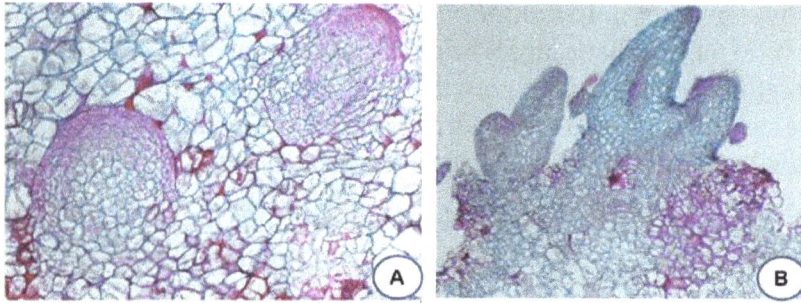

A. Section of a 2-week-old regenerating callus from endosperm of *A. indica* showing endogenous meristematic pockets. **B.** Section of a 4-week-old regenerating callus showing distinct shoot buds differentiated from peripheral vascularized nodule. One of the buds showing glands.

Cytology

The endosperm tissue often shows a high degree of chromosomal variation and polyploidy. Mitotic irregularities, chromosome bridges and laggards are other important characteristic features of endosperm tissue. Some reports suggest that the cells of endosperm cultures showed ploidy higher than 3n as in the case of *Croton* (Bhojwani and Johri 1971), *Jatropha* (Srivastava 1971a) and *Lolium* (Norstog et al. 1969). Cytological observations of the endosperm callus, derived from *Dendrophthoe falcata*, *Taxillus cuneatus* and *Taxillux vestitus*, showed diploid (2n=18) and triploid (3n=27) chromosomes (Johri and Nag 1974).

In addition to the cytological observations of endosperm callus, the chromosomal analysis of the regenerated plantlets were also studied in a number of systems. In *Juglans regia* two plants of endosperm origin were analysed for ploidy determination and both the plants showed triploid (3n=3x=48) number of chromosomes (Tulecke et al. 1988). In *Citrus* stability of the ploidy level and chromosome number were observed all through the regeneration process and triploid (2n=3x=27) plantlets were recovered (Gmitter et al. 1990). In *Mallotus philippensis* the squash preparation of root tip cells of the regenerated plants invariably showed triploid chromosome number (3n=3x=33) (Sehgal and Abbas 1996). The triploid nature of the endosperm-derived plants was determined by Feulgen cytophotometry in *Acacia nilotica* (Garg et al. 1996).

In Mulberry (*Morus alba*), 7-month-old plants of endosperm origin were utilized for ploidy determination. All the ten plants analysed cytologically showed triploid number of chromosome (2n=3x=42) (Thomas et al. 2000). The ploidy determination of 20 plants of *Azadirachta indica* , regenerated from endosperm calli, showed that 66% of the plants had triploid number of chromosomes (2n=3x=36) and the rest 34% were diploids (2n=2x=24) (Chaturvedi et al. 2003). In *Actinidia deliciosa* three different ploidy levels viz., 3C, 6C and 9C were observed in cells of endosperm derived callus analyzed by flow cytometry. The analysis of the leaves of endosperm derived plants showed 45.7% fluorescence intensity peaks corresponding exactly to 3C whereas 42.2% exhibited peaks of fluorescence intensity representing C-values between 2C and 4C.

Only 8.4% of the samples indicated 2C DNA content, and one sample showed 6C DNA content (Goralski et al. 2005).

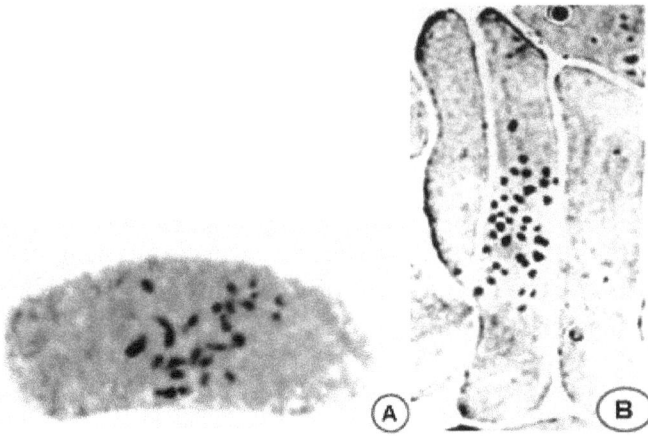

Cells from the root-tips of shoots of endosperm origin: **A.** showing diploid number of chromosomes (2n=2x=24), **B.** triploid number of chromosomes (2n=3x=36)

Concluding Remarks

Like all other plant parts, endosperm can also respond the same way under in vitro conditions irrespective of their genetic constitutions. Hence, it helped in changing the misconception that endosperm being a "dead tissue" has now been contradicted by several reports suggesting the full plant regeneration from endosperm. Despite the success of plant regeneration from endosperm cultures in a number of systems, this protocol for production of triploid plants remained unutilised mostly. It may be basically due to the difficulty in obtaining organogenic callus from the mature or immature endosperms. The ploidy variation exhibited by endosperm derived plantlets is another difficulty which limits this technique. More efforts should be focussed on endosperm regeneration from plants where seedlessness is employed to improve the quality of fruits and plants that has economically useful vegetative parts.

References

- J. Mark Cock; Sheila McCormick (July 2001). "A Large Family of Genes That Share Homology with CLAVATA3". Plant Physiology. 126: 939–942. doi:10.1104/pp.126.3.939. PMC 1540125. PMID 11457943

- "Branching out: new class of plant hormones inhibits branch formation". Nature. 455 (7210). 2008-09-11. Retrieved 2009-04-30

- Valster, A. H.; et al. (2000). "Plant GTPases: the Rhos in bloom". Trends in Cell Biology. 10 (4): 141–146. doi:10.1016/s0962-8924(00)01728-1

- Galun, Esra (2007). Plant Patterning: Structural and Molecular Genetic Aspects. World Scientific Publishing Company. p. 333. ISBN 9789812704085

- Mayer, K. F. X; et al. (1998). "Role of WUSCHEL in Regulating Stem Cell Fate in the Arabidopsis Shoot Meristem". Cell. 95: 805–815. doi:10.1016/S0092-8674(00)81703-1. PMID 9865698

Plant Cell Culture and Protoplast Fusion

The aggregation found in cells while suspended in a liquid medium is known as cell suspension culture. The common manner of cell suspension is by transferring a friable callus which helps in the dispersing of cells and the eliminating of large callus pieces. The topics discussed in the chapter are of great importance to broaden the existing knowledge on cell suspension cultures.

Cell Culture

In spite of advances in synthetic organic chemistry, chemical synthesis of several compounds is not yet feasible due to their complex structures. The plants still contribute significantly to the bulk of the market products, such as secondary metabolites. The major limitations to the commercial use of potential metabolites is their very scarce supply from the field grown plants due to their seasonal growth, genetic, geographical and climatic variations, and insect and pathogen attack. The environmental fluxes cause alterations in type and quantity of metabolites produced. In this context, the cell culture technique is complimentary and may provide competitive metabolite production systems when compared to whole plant extraction. Since the plant cells are totipotent in nature, consequently, the cultured cells will also contain the genetic information for the production of therapeutic compounds with added potential of increasing yield by explants selection and manipulation of culture conditions. The cell cultures have several advantages over conventional isolation of metabolites from the intact plants, such as stable supply, freedom from disease and vagaries of climates, closer relationship between supply and demand, and growth of large amount of plant tissues in minimal space. Such an alternative method has considerable implication as it would reduce the pressure on natural population and, thus, may prevent the plant from becoming endangered. It may also help to steer clear of contaminated pharmaceutical raw material.

Also, studies on secondary metabolites require an in-depth understanding of biosynthetic pathways which is often difficult to conduct in whole plants because the biosynthetic activities may only be expressed in particular cell types within a specific plant organ or at a certain time of season. Cell cultures have a higher rate of metabolism than intact differentiated plants because the initiation of cell growth in culture leads to fast proliferation of cell mass and to a condensed biosynthetic cycle. As a result, secondary metabolite production can take place within a short cultivation time (about 2-4 weeks) with an added advantage of tunability. These cell cultures can be employed in scaled-up operation, for isolation of desirable compounds in bulk.

History and Evolution

Haberlandt attempted to cultivate isolated plant cells, but cell division was never observed in these cultures. In the 1930s the first *in vitro* cultures were established, and followed by the development of culture media and culture methods. Twenty-five years ago, the prospect of the use of plant cell cultures for metabolite production is not imaginable. The low yields of metabolite production in suspension cultures were the bottlenecks for commercialization. In such early efforts, plant cells in culture were treated in direct analogy to microbial systems, with little knowledge of plant cell physiology and biochemistry. In 1982, at least 30 compounds were known to accumulate in plant culture systems in concentrations equal to or higher than that of the plant. Concerning the production of natural metabolites, some *in vitro* systems exist that allow the large-scale production of economically important metabolites. A survey of the historical milestones in plant cell cultures is given by Schmauder and Doebel in 1990. Strategies to optimize growth and product formation began to develop separately during the period between 1975 and 1985. A combination of strategies for yield improvement resulted in the first commercial plant cell process.

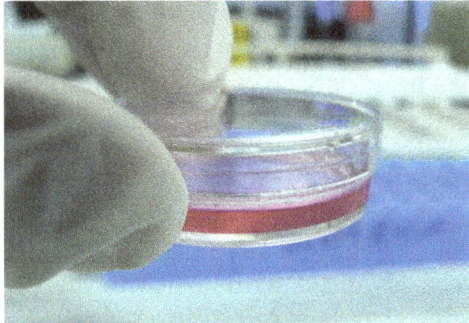

Cell culture in a special tissue culture dish

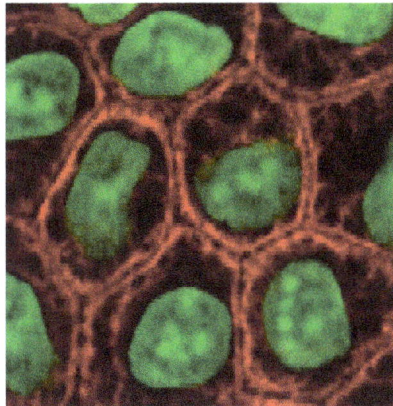

Epithelial cells in culture, stained for keratin (red) and DNA (green)

Cell culture is the process by which cells are grown under controlled conditions, generally outside of their natural environment. Cell culture conditions can vary for each cell type, but artificial environments consist of a suitable vessel with substrate or medium that supplies the essential nutrients (amino acids, carbohydrates, vitamins, minerals), growth factors, hormones, and gases (CO_2, O_2), and regulates the physio-chemical environment (pH buffer,osmotic pressure, temperature). Most cells require a surface or an artificial substrate (adherent or monolayer culture) whereas others can be grown free floating in culture medium (suspension culture).

In practice, the term "cell culture" now refers to the culturing of cells derived from multicellular eukaryotes, especially animal cells, in contrast with other types of culture that also grow cells, such as plant tissue culture, fungal culture, and microbiological culture (of microbes). The historical development and methods of cell culture are closely interrelated to those of tissue culture and organ culture. Viral culture is also related, with cells as hosts for the viruses.

The laboratory technique of maintaining live cell lines (a population of cells descended from a single cell and containing the same genetic makeup) separated from their original tissue source became more robust in the middle 20th century.

History

The 19th-century English physiologist Sydney Ringer developed salt solutions containing the chlorides of sodium, potassium, calcium and magnesium suitable for maintaining the beating of an isolated animal heart outside of the body. In 1885, Wilhelm Roux removed a portion of the medullary plate of an embryonic chicken and maintained it in a warm saline solution for several days, establishing the principle of tissue culture.Ross Granville Harrison, working at Johns Hopkins Medical School and then at Yale University, published results of his experiments from 1907 to 1910, establishing the methodology of tissue culture.

Cell culture techniques were advanced significantly in the 1940s and 1950s to support research in virology. Growing viruses in cell cultures allowed preparation of purified viruses for the manufacture of vaccines. The injectable polio vaccine developed by Jonas Salk was one of the first products mass-produced using cell culture techniques. This vaccine was made possible by the cell culture research of John Franklin Enders, Thomas Huckle Weller, and Frederick Chapman Robbins, who were awarded a Nobel Prize for their discovery of a method of growing the virus in monkey kidney cell cultures.

Concepts in Mammalian Cell Culture

Isolation of Cells

Cells can be isolated from tissues for *ex vivo* culture in several ways. Cells can be easily purified from blood; however, only the white cells are capable of growth in culture. Mononuclear cells can be released from soft tissues by enzymatic digestion with enzymes such as collagenase, trypsin, or pronase, which break down the extracellular matrix. Alternatively, pieces of tissue can be placed in growth media, and the cells that grow out are available for culture. This method is known as explant culture.

Cells that are cultured directly from a subject are known as primary cells. With the exception of some derived from tumors, most primary cell cultures have limited lifespan.

An established or immortalized cell line has acquired the ability to proliferate indefinitely either through random mutation or deliberate modification, such as artificial expression of the telomerasegene. Numerous cell lines are well established as representative of particular cell types.

Maintaining Cells in Culture

For the majority of isolated primary cells, they undergo the process of senescence and stop dividing after a certain number of population doublings while generally retaining their viability (described as the Hayflick limit).

Cells are grown and maintained at an appropriate temperature and gas mixture (typically, 37 °C, 5% CO_2 for mammalian cells) in a cell incubator. Culture conditions vary widely for each cell type, and variation of conditions for a particular cell type can result in different phenotypes.

A bottle of DMEM cell culture medium

Aside from temperature and gas mixture, the most commonly varied factor in culture systems is the cell growth medium. Recipes for growth media can vary in pH, glucose concentration, growth factors, and the presence of other nutrients. The growth factors used to supplement media are often derived from the serum of animal blood, such as fetal bovine serum (FBS), bovine calf serum, equine serum, and porcine serum. One complication of these blood-derived ingredients is the potential for contamination of the culture with viruses or prions, particularly in medical biotechnology applications. Current practice is to minimize or eliminate the use of these ingredients wherever possible and use human platelet lysate (hPL). This eliminates the worry of cross-species contamination when using FBS with human cells. hPL has emerged as a safe and reliable alternative as a direct replacement for FBS or other animal serum. In addition, chemically defined media can be used to eliminate any serum trace (human or animal), but this cannot always be accomplished with different cell types. Alternative strategies involve sourcing the animal blood from countries with minimum BSE/TSE risk, such as The United States, Australia and New Zealand, and using purified nutrient concentrates derived from serum in place of whole animal serum for cell culture.

Plating density (number of cells per volume of culture medium) plays a critical role for some cell types. For example, a lower plating density makes granulosa cells exhibit estrogen production, while a higher plating density makes them appear as progesterone-producing theca lutein cells.

Cells can be grown either in suspension or adherent cultures. Some cells naturally live in suspension, without being attached to a surface, such as cells that exist in the bloodstream. There are also cell lines that have been modified to be able to survive in suspension cultures so they can be grown to a higher density than adherent conditions would allow. Adherent cells require a surface, such as tissue culture plastic or microcarrier, which may be coated with extracellular matrix (such as collagen and laminin) components to increase adhesion properties and provide other signals needed for growth and differentiation. Most cells derived from solid tissues are adherent. Another type of adherent culture is organotypic culture, which involves growing cells in a three-dimensional (3-D) environment as opposed to two-dimensional culture dishes. This 3D culture system is biochemically and physiologically more similar to *in vivo* tissue, but is technically challenging to maintain because of many factors (e.g. diffusion).

Components of Cell Culture Media

Component	Function
Carbon source (glucose/ glutamine)	Source of energy
Amino acid	Building blocks of protein
Vitamins	Promote cell survival and growth
Balanced salt solution	An isotonic mixture of ions to maintain optimum osmotic pressure within the cells and provide essential metal ions to act as cofactors for enzymatic reactions, cell adhesion etc.
Phenol red dye	pH indicator. The color of phenol red changes from orange/red at pH 7-7.4 to yellow at acidic (lower) pH and purple at basic (higher) pH.
Bicarbonate /HEPES buffer	It is used to maintain a balanced pH in the media

Growth Conditions

Parameter	
Temperature	37°C
CO2	5%
Humidity	95%

Cell line Cross-contamination

Cell line cross-contamination can be a problem for scientists working with cultured cells. Studies suggest anywhere from 15–20% of the time, cells used in experiments have been misidentified or contaminated with another cell line. Problems with cell line cross-contamination have even been detected in lines from the NCI-60 panel, which are used routinely for drug-screening studies. Major cell line repositories, including the American Type Culture Collection (ATCC), the European Collection of Cell Cultures (ECACC) and the German Collection of Microorganisms and Cell Cultures (DSMZ), have received cell line submissions from researchers that were misidentified by them. Such contamination poses a problem for the quality of research produced using cell culture lines, and the major repositories are now authenticating all cell line submissions. ATCC uses short tandem repeat (STR) DNA fingerprinting to authenticate its cell lines.

To address this problem of cell line cross-contamination, researchers are encouraged to authenticate their cell lines at an early passage to establish the identity of the cell line. Authentication should be repeated before freezing cell line stocks, every two months during active culturing and before any publication of research data generated using the cell lines. Many methods are used to identify cell lines, including isoenzyme analysis, human lymphocyte antigen (HLA) typing, chromosomal analysis, karyotyping, morphology and STR analysis.

One significant cell-line cross contaminant is the immortal HeLa cell line.

Other Technical Issues

As cells generally continue to divide in culture, they generally grow to fill the available area or volume. This can generate several issues:

- Nutrient depletion in the growth media

- Changes in pH of the growth media

- Accumulation of apoptotic/necrotic (dead) cells

- Cell-to-cell contact can stimulate cell cycle arrest, causing cells to stop dividing, known as contact inhibition.

- Cell-to-cell contact can stimulate cellular differentiation.
- Genetic and epigenetic alterations, with a natural selection of the altered cells potentially leading to overgrowth of abnormal, culture-adapted cells with decreased differentiation and increased proliferative capacity.

Manipulation of Cultured Cells

Among the common manipulations carried out on culture cells are media changes, passaging cells, and transfecting cells. These are generally performed using tissue culture methods that rely on aseptic technique. Aseptic technique aims to avoid contamination with bacteria, yeast, or other cell lines. Manipulations are typically carried out in a biosafety hood or laminar flow cabinet to exclude contaminating micro-organisms. Antibiotics (e.g. penicillin and streptomycin) and antifungals (e.g.amphotericin B) can also be added to the growth media.

As cells undergo metabolic processes, acid is produced and the pH decreases. Often, a pH indicator is added to the medium to measure nutrient depletion.

Media Changes

In the case of adherent cultures, the media can be removed directly by aspiration, and then is replaced. Media changes in non-adherent cultures involve centrifuging the culture and resuspending the cells in fresh media.

Passaging Cells

Passaging (also known as subculture or splitting cells) involves transferring a small number of cells into a new vessel. Cells can be cultured for a longer time if they are split regularly, as it avoids the senescence associated with prolonged high cell density. Suspension cultures are easily passaged with a small amount of culture containing a few cells diluted in a larger volume of fresh media. For adherent cultures, cells first need to be detached; this is commonly done with a mixture of trypsin-EDTA; however, other enzyme mixes are now available for this purpose. A small number of detached cells can then be used to seed a new culture. Some cell cultures, such as RAW cells are mechanically scraped from the surface of their vessel with rubber scrapers.

Transfection and Transduction

Another common method for manipulating cells involves the introduction of foreign DNA by transfection. This is often performed to cause cells to express a gene of interest. More recently, the transfection of RNAi constructs have been realized as a convenient mechanism for suppressing the expression of a particular gene/protein. DNA can also be inserted into cells using viruses, in methods referred to as transduction, infection or transformation. Viruses, as parasitic agents, are well suited to introducing DNA into cells, as this is a part of their normal course of reproduction.

Established Human Cell Lines

Cell lines that originate with humans have been somewhat controversial in bioethics, as they may outlive their parent organism and later be used in the discovery of lucrative medical treatments. In the pioneering decision in this area, the Supreme Court of California held in *Moore v. Regents of the University of California* that human patients have no property rights in cell lines derived from organs removed with their consent.

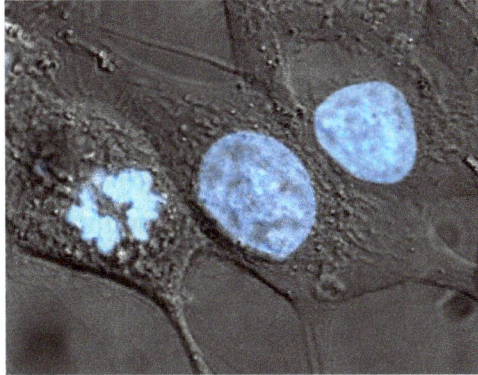

Cultured HeLa cells have been stained with Hoechst turning their nuclei blue, and are one of the earliest human cell lines descended from Henrietta Lacks, who died of cervical cancer from which these cells originated.

It is possible to fuse normal cells with an immortalised cell line. This method is used to produce monoclonal antibodies. In brief, lymphocytes isolated from the spleen (or possibly blood) of an immunised animal are combined with an immortal myeloma cell line (B cell lineage) to produce a hybridoma which has the antibody specificity of the primary lymphocyte and the immortality of the myeloma. Selective growth medium (HA or HAT) is used to select against unfused myeloma cells; primary lymphoctyes die quickly in culture and only the fused cells survive. These are screened for production of the required antibody, generally in pools to start with and then after single cloning.

Cell Strains

A cell strain is derived either from a primary culture or a cell line by the selection or cloning of cells having specific properties or characteristics which must be defined. Cell strains are cells that have been adapted to culture but, unlike cell lines, have a finite division potential. Non-immortalized cells stop dividing after 40 to 60 population doublings and, after this, they lose their ability to proliferate (a genetically determined event known as senescence).

Applications of Cell Culture

Mass culture of animal cell lines is fundamental to the manufacture of viral vaccines and other products of biotechnology. Culture of human stem cells is used to expand the number of cells and differentiate the cells into various somatic cell types for transplantation. Stem cell culture is also used to harvest the molecules and exosomes that the stem cells release for the purposes of therapeutic development.

Biological products produced by recombinant DNA (rDNA) technology in animal cell cultures include enzymes, synthetic hormones, immunobiologicals (monoclonal antibodies, interleukins, lymphokines), and anticancer agents. Although many simpler proteins can be produced using rDNA in bacterial cultures, more complex proteins that are glycosylated (carbohydrate-modified) currently must be made in animal cells. An important example of such a complex protein is the hormone erythropoietin. The cost of growing mammalian cell cultures is high, so research is underway to produce such complex proteins in insect cells or in higher plants, use of single embryonic cell and somatic embryos as a source for direct gene transfer via particle bombardment, transit gene expression and confocal microscopy observation is one of its applications. It also offers to confirm single cell origin of somatic embryos and the asymmetry of the first cell division, which starts the process.

Cell culture is also a key technique for cellular agriculture, which aims to provide both new products and new ways of producing existing agricultural products like milk, (cultured) meat, fragrances, and rhino horn from cells and microorganisms. It is therefore considered one means of achieving animal-free agriculture.

Cell Culture in Two Dimensions

Research in tissue engineering, stem cells and molecular biology primarily involves cultures of cells on flat plastic dishes. This technique is known as two-dimensional (2D) cell culture, and was first developed by Wilhelm Roux who, in 1885, removed a portion of the medullary plate of an embryonic chicken and maintained it in warm saline for several days on a flat glass plate. From the advance of polymer technology arose today's standard plastic dish for 2D cell culture, commonly known as the Petri dish. Julius Richard Petri, a German bacteriologist, is generally credited with this invention while working as an assistant to Robert Koch. Various researchers today also utilize culturing laboratory flasks, conicals, and even disposable bags like those used in single-use bioreactors.

Aside from Petri dishes, scientists have long been growing cells within biologically derived matrices such as collagen or fibrin, and more recently, on synthetic hydrogels such as polyacrylamide or PEG. They do this in order to elicit phenotypes that are not expressed on conventionally rigid substrates. There is growing interest in controlling matrix stiffness, a concept that has led to discoveries in fields such as:

- Stem cell self-renewal
- Lineage specification
- Cancer cell phenotype
- Fibrosis
- Hepatocyte function
- Mechanosensing

Cell Culture in Three Dimensions

Cell culture in three dimensions has been touted as "Biology's New Dimension". At present, the practice of cell culture remains based on varying combinations of single or multiple cell structures in 2D. Currently, there is an increase in use of 3D cell cultures in research areas including drug discovery, cancer biology, regenerative medicine and basic life science research. There are a variety of platforms used to facilitate the growth of three-dimensional cellular structures including scaffold systems such as hydrogel matrices and solid scaffolds, and scaffold-free systems such as low-adhesion plates, nanoparticle facilitated magnetic levitation, and hanging drop plates.

3D Cell Culture in Hydrogels

As the natural extracellular matrix (ECM) is important in the survival, proliferation, differentiation and migration of cells, different hydrogel culture matrices mimicking natural ECM structure are seen as potential approaches to in vivo –like cell culturing. Hydrogels are composed of interconnected pores with high water retention, which enables efficient transport of substances such as nutrients and gases. Several different types of hydrogels from natural and synthetic materials are available for 3D cell culture, including animal ECM extract hydrogels, protein hydrogels, peptide hydrogels, polymer hydrogels, and wood-based nanocellulose hydrogel.

3D Cell Culturing by Magnetic Levitation

The 3D Cell Culturing by Magnetic Levitation method (MLM) is the application of growing 3D tissue by inducing cells treated with magnetic nanoparticle assemblies in spatially varying magnetic fields using neodymium magnetic drivers and promoting cell to cell interactions by levitating the cells up to the air/liquid interface of a standard petri dish. The magnetic nanoparticle assemblies consist of magnetic iron oxide nanoparticles, gold nanoparticles, and the polymer polylysine. 3D cell culturing is scalable, with the capability for culturing 500 cells to millions of cells or from single dish to high-throughput low volume systems.

Tissue Culture and Engineering

Cell culture is a fundamental component of tissue culture and tissue engineering, as it establishes the basics of growing and maintaining cells *in vitro*. The major application of human cell culture is in stem cell industry, where mesenchymal stem cells can be cultured and cryopreserved for future use. Tissue engineering potentially offers dramatic improvements in low cost medical care for hundreds of thousands of patients annually.

Vaccines

Vaccines for polio, measles, mumps, rubella, and chickenpox are currently made in cell cultures. Due to the H5N1pandemic threat, research into using cell culture for influenza vaccines is being funded by the United States government. Novel ideas in the field include recombinant DNA-based vaccines, such as one made using human adenovirus (a common cold virus) as a vector, and novel adjuvants.

Culture of Non-mammalian Cells

Plant Cell Culture Methods

Plant cell cultures are typically grown as cell suspension cultures in a liquid medium or as callus cultures on a solid medium. The culturing of undifferentiated plant cells and calli requires the proper balance of the plant growth hormones auxin and cytokinin.

Insect Cell Culture

Cells derived from Drosophila melanogaster (most prominently, Schneider 2 cells) can be used for experiments which may be hard to do on live flies or larvae, such as biochemical studies or studies using siRNA. Cell lines derived from the army worm *Spodoptera frugiperda*, including Sf9 and Sf21, and from the cabbage looper *Trichoplusia ni*, High Five cells, are commonly used for expression of recombinant proteins using baculovirus.

Bacterial and Yeast Culture Methods

For bacteria and yeasts, small quantities of cells are usually grown on a solid support that contains nutrients embedded in it, usually a gel such as agar, while large-scale cultures are grown with the cells suspended in a nutrient broth.

Viral Culture Methods

The culture of viruses requires the culture of cells of mammalian, plant, fungal or bacterial origin as hosts for the growth and replication of the virus. Whole wild type viruses, recombinant viruses or viral products may be generated in cell types other than their natural hosts under the right conditions. Depending on the species of the virus, infection and viral replication may result in host cell lysis and formation of a viral plaque.

Common Cell Lines

Human cell lines

- DU145 (prostate cancer)
- H295R (adrenocortical cancer)
- HeLa (cervical cancer)
- KBM-7 (chronic myelogenous leukemia)
- LNCaP (prostate cancer)
- MCF-7 (breast cancer)

- MDA-MB-468 (breast cancer)

- PC3 (prostate cancer)

- SaOS-2 (bone cancer)

- SH-SY5Y (neuroblastoma, cloned from a myeloma)

- T47D (breast cancer)

- THP-1 (acute myeloidleukemia)

- U87 (glioblastoma)

- National Cancer Institute's 60 cancer cell line panel (NCI60)

Primate cell lines

- Vero (African green monkey *Chlorocebus* kidney epithelial cell line)

Mouse cell lines

- MC3T3 (embryonic calvarium)

Rat tumor cell lines

- GH3 (pituitary tumor)

- PC12 (pheochromocytoma)

Plant cell lines

- Tobacco BY-2 cells (kept as cell suspension culture, they are model system of plant cell)

Other species cell lines

- Dog MDCK kidney epithelial

- Xenopus A6 kidney epithelial

- ZebrafishAB9

Immortalised Cell Line

An immortalised cell line is a population of cells from a multicellularorganism which would normally not proliferate indefinitely but, due to mutation, have evaded normal cellular senescence and instead can keep undergoing division. The cells can therefore be grown for prolonged periods *in vitro*. The mutations required for immortality can occur naturally or be intentionally induced for experimental purposes. Immortal cell lines are a very important tool for research into the biochemistry and cell biology of multicellular organisms. Immortalised cell lines have also found uses in biotechnology.

An immortalised cell line should not be confused with stem cells, which can also divide indefinitely, but form a normal part of the development of a multicellular organism.

Relation to Natural Biology and Pathology

There are various immortal cell lines. Some of them are normal cell lines - e.g. derived from stem cells. Other immortalised cell lines are the *in vitro* equivalent of cancerous cells. Cancer occurs when a somatic cell which normally cannot divide undergoes mutations which cause de-regulation of the normal cell cycle controls leading to uncontrolled proliferation. Immortalised cell lines have undergone similar mutations allowing a cell type which would normally not be able to divide to be proliferated *in vitro*. The origins of some immortal cell lines, for example HeLa human cells, are from naturally occurring cancers.

Role and Uses

Immortalised cell lines are widely used as a simple model for more complex biological systems, for example for the analysis of the biochemistry and cell biology of mammalian (including human) cells. The main advantage of using an immortal cell line for research is its immortality; the cells can be grown indefinitely in culture. This simplifies analysis of the biology of cells which may otherwise have a limited lifetime.

Immortalised cell lines can also be cloned giving rise to a clonal population which can, in turn, be propagated indefinitely. This allows an analysis to be repeated many times on genetically identical cells which is desirable for repeatable scientific experiments. The alternative, performing an analysis on primary cells from multiple tissue donors, does not have this advantage.

Immortalised cell lines find use in biotechnology where they are a cost-effective way of growing cells similar to those found in a multicellular organism *in vitro*. The cells are used for a wide variety of purposes, from testing toxicity of compounds or drugs to production of eukaryotic proteins.

Limitations

Changes from Nonimmortal Origins

While immortalised cell lines often originate from a well-known tissue type they have undergone significant mutations to become immortal. This can alter the biology of the cell and must be taken into consideration in any analysis. Further, cell lines can change genetically over multiple passages, leading to phenotypic differences among isolates and potentially different experimental results depending on when and with what strain isolate an experiment is conducted with.

Contamination with Other Cells

Many cell lines that are widely used for biomedical research have been contaminated and overgrown by other, more aggressive cells. For example, supposed thyroid lines were actually melanoma cells, supposed prostate tissue was actually bladder cancer, and supposed normal uterine cultures were actually breast cancer.

Methods of Generation

There are several methods for generating immortalised cell lines:

1. Isolation from a naturally occurring cancer. This is the original method for generating an immortalised cell line. Major examples include human HeLa cells that were obtained from a cervical cancer, mouse Raw 264.7 cells that were subjected to mutagenesis and then selected for cells which are able to undergo division.

2. Introduction of a viral gene that partially deregulates the cell cycle (e.g., the adenovirus type 5 E1 gene was used to immortalize the HEK 293 cell line).

3. Artificial expression of key proteins required for immortality, for example telomerase which prevents degradation of chromosome ends during DNA replication in eukaryotes

4. Hybridoma technology, specifically used for the generation of immortalised antibody-producing B cell lines, where an antibody-producing B cell is fused with a myeloma (B cell cancer) cell.

Examples

There are several examples of immortalised cell lines, each with different properties. Most immortalised cell lines are classified by the cell type they originated from or are most similar to biologically:

- 3T3 cells – a mouse fibroblast cell line derived from a spontaneous mutation in cultured mouse embryo tissue

- A549 cells – derived from a cancer patient lung tumor

- F11 cells – a line of neurons from the dorsal root ganglia of rats

- HeLa cells – a widely used human cell line isolated from cervical cancer patient, Henrietta Lacks

- HEK 293 cells – derived from human fetal cells

- Jurkat cells – a human T lymphocyte cell line isolated from a case of leukemia

- Vero cells – a monkey kidney cell line that arose by spontaneous immortalisation

Initiation and Establishment of Cell Suspension Cultures

Callus Cultures

When an organ of a plant is damaged a wound repair response is induced to bring about the repair of the damaged portion. This response is associated with the induction of division in the undamaged cells adjacent to the lesion, thus sealing of the wound. If, however, wounding is followed by the aseptic culture of the damaged region on a defined medium, the initial cell division response can be stimulated and induced to continue indefinitely through the exogenous influence of the chemical constitution of the culture medium. The result is a continually-dividing mass of cells without any significant differentiation and organization and this proliferated mass of cell aggregate is called callus. The first step to establish cell suspension cultures is to raise callus from any explants of the plant. To maximize the production of a particular compound, it is desirable to initiate the callus from the plant part that is known to be high producer.

Calli are generally grown on medium solidified with gelling agents like, agar, gelrite, agarose, in Petri-dishes, glass test-tubes or extra-wide necked Erlenmeyer flasks. In morphological terms it can vary extensively, ranging from being very hard/compact and green or light green in color, where the cells have extensive and strong cell to cell contact, to being 'friable' where the callus consists of small, disintegrating aggregates of poorly-associated cells and has brownish or creamy appearance. Friable callus is most demanded since it shows fast and uniform growth of cells and is highly suitable to initiate cell suspension cultures. Callus morphology is explants and species dependent but can be altered by the modification of plant growth regulators in the medium.

The callus cultures shows inherent degree of heterogeneity and this may be due to their size and nature, unidirectional supply of nutrient medium (below the callus) and gases and light (predominantly from above). The heterogeneity may be disadvantageous in uniform production of cell biomass but may be useful in the developmental responses of the callus like, shoot regeneration.

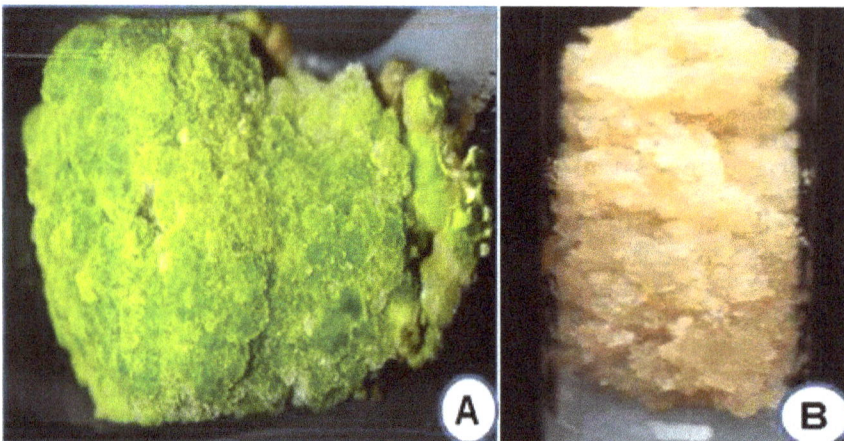

Callus cultures- **A.** Hard and compact callus; **B.** Friable and brown callus

Cell Suspension Culture

Characteristics: A suspension culture is developed by transferring the relatively friable portion of a callus as in figure, into liquid medium and is maintained under suitable conditions of aeration, agitation, light, temperature and other physical parameters. The increased cell dissociation means increased culture uniformity. Plant cells are significantly larger and slower growing cells than most microbial organisms. They mostly resemble to parenchymatous cells in having relatively large vacuoles, a thin layer of cytoplasm and thin, rounded cell walls. The species/genotypes and medium composition used can influence *in vitro* cell morphology and different cell types with different morphological/ physiological properties can co-exist within a single culture.

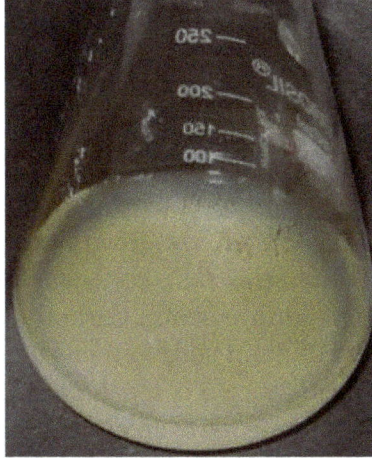
Cells in liquid medium showing fine suspension of cells

Cell growth: The most commonly used cell suspensions are of the closed (or batch) type where the cells are grown in fixed volume of liquid medium and which are routinely maintained through the transfer of a portion (ca 10%) of a fully-grown culture to fresh medium at regular intervals. The growth curve of a cell suspension culture has a characteristic shape consisting of four essential stages- an initial lag phase, an exponential phase, stationary phase and death phase. The duration of each phase is dependent on the species or genotype selected, explant used, culture medium and subculture regime. The lag phase is shortened when relatively large inocula are used although paradoxically, growth terminates earlier and overall biomass production is reduced.

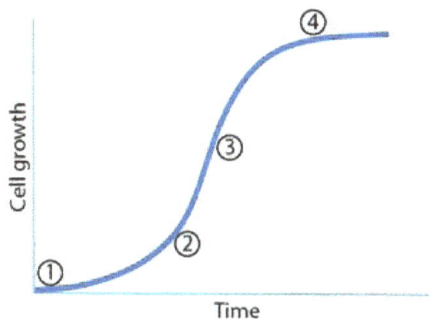
Growth curve for plant cell suspension grown in closed system. The four different growth phases are labeled: (1) Lag phase, (2) Exponential phase, (3) Linear phase, (4) Stationary phase.

Aggregation: Due to larger size of a plant cell, it is capable of withstanding tensile strain but is sensitive to shear stress. Aggregation is common, largely due to failure of the cells to separate after division. The secretion of extracellular polysaccharides, particularly in the later stages of growth, may further contribute to increased adhesion. This tendency of plant cells to grow in clumps results in sedimentation, insufficient mixing and diffusion-limited biochemical reaction. Even the fine suspension culture consists of micro-to sub-macroscopic colonies made up of around 5-200 cells and such degree of cell aggregation is acceptable. Cultures consisting of larger aggregates like, 0.5-1.0 mm in diameter, are more readily attainable, grow perfectly well and depending on the aim of the research are often sufficient to meet all requirements. This so called cell-cell contact is desirable for the biosynthesis of many secondary metabolites by the plant cells in suspension cultures. Therefore, controlled aggregation of plant cells may be of interest from process engineering point of view.

Oxygen and aeration effect: Oxygen requirements of plant cells are comparatively lower than that of microbial cells due to their low growth rates. In some cases high oxygen concentration is even toxic to the cell's metabolic activities and may strip nutrients such as carbon dioxide from the culture broth. Carbon dioxide is often considered as essential nutrient in the culture of plant cells and has a positive effect on cell growth. Moderate shaking speed like, 90-120 rpm is ideal for standard aeration. As the plant cells are shear sensitive and the immediate effects of high agitation are the cell damage, reduction in cell viability, release of intracellular compounds while low agitation (<90rpm) results in cell aggregation.

Plant Cell Cultures Vs. Microbial Cultures

Although basic equipment-and process-related requirements are same for both plant and microbial cell suspension cultures but some of the features suitable for microbial cells cannot be used for plant cells due to prominent differences in the nature and growth pattern of the two types of cells:

1. Plant cells are sensitive to shear stresses because of the relatively inflexible cell wall and their large size (50-100 μm) compared to microbial cells (2-10 μm).

2. Plant cell cultures show relatively long growth cycles. Typical specific growth rates (increase in cell mass per unit time) may range from $0.12d^{-1}$ to $0.05d^{-1}$; thus, the typical doubling time of plant cell cultures is measured in days, as compared to hours for microbial systems.

3. In plant cells, the vacuole is usually the site of product accumulation and extracellular product secretion is rare. While in microbial cells product accumulation often extracellular.

4. The plant cells mostly grow as aggregates while microbial cells grow as single cells.

5. The plant cells in suspension cultures often undergo spontaneous genetic variation in terms of accumulation of secondary metabolites, which leads to heterogeneous population of cells in a suspension culture. Compared to this microbial cells are genetically stable.

Experiment

Aim: To raise cell suspension culture of *Lantana camara* .

Equipments: Autoclave, pH-meter, Magnetic stirrers, Magnetic beads, Weighing balance, Laminar-air-flow, Microsopes, Incubator shaker.

Materials Required : Salts and vitamins of Murashige and Skoog's (MS; 1962), sucrose, agar, conical flasks, measuring cylinders and beakers of various sizes. Reagent glass bottles for storage, spatula, tissue rolls, distilled water. Cotton plugs, aluminium foils, muslin cloth, scissor, media stocks, 1N NaOH, 1N HCl, myo-inositol. Autoclavable polybags, rubber bands. Borosil glass test-tubes (150mm x 25mm without rim) and Erlenmeyer flasks. Black markers, micropipette, micropipette-tips, test-tube stands, autoclavable baskets, plastic trays, cork borer, Leaves from healthy *L. camara* plants bearing pink-yellow colored flowers.

Plant Growth regulators (Sigma):

Benzyl amino purine (BAP), Naphthalene acetic acid (NAA) , 2-4-dichlorophenoxy-acetic acid.

Methods

i. Initiation and establishment of leaf-disc cultures

Leaves were washed with 1% Tween-20 (Merck, India) for 15 min, followed by three rinses in sterile distilled water (SDW). Thereafter, the remaining steps were carried out inside the laminar-air-flow cabinet (Saveer Biotech, India). Leaves were surface sterilized with 0.1% mercuric chloride solution ($HgCl_2$) for 10 min and rinsed thrice with SDW. Leaf-disc explants were prepared by punching the sterilized leaves with 5 mm sized cork-borer before being cultured with the adaxial side in contact with the media.

ii. Establishment of callus cultures

Leaf disc explants were incubated on MS (with 3% sucrose) basal medium consisting of macro and microsalts, vitamins, iron, 100 mgl^{-1} myoinositol and solidified with 0.8% agar (HiMedia, India). The medium was enriched with combination of cytokinin and auxins, BAP (5μM) + NAA (1 μM) + 2,4-D (1 μM) for induction and multiplication of callus from leaf-disc explants. After adjusting the pH to 5.8, 20 ml of the medium was dispensed into each 150 x 25 mm Borosil rimless glass tubes. The culture tubes were plugged with nonabsorbent cotton wrapped in cheesecloth and autoclaved at 121°C at

15 psi for 15 min. All the cultures were maintained in diffuse light (1000-1600 lux) and 16 h photoperiod at $25 \pm 2°C$ and 50-60% relative humidity. Observations were recorded at weekly intervals. After callus induction, the biomass was multiplied constantly by inoculating 0.2 g of calli onto fresh medium at every 4-week intervals.

iii. Establishment of cell suspension cultures

After 7-8 passages of callus subcultures, healthy, green, friable and soft calli maintained on responding semi-solid medium were utilized to establish suspension cultures. Cultures were initiated in Erlenmeyer flasks of 250 ml capacity, containing 50 ml of liquid medium and inoculated with 0.2 g of fresh calli. The cultures were incubated in an orbital shaker under shaking conditions at 120 rpm at $25°C \pm 2°C$ and maintained in diffuse light (1000-1600 lux) with 16 h photoperiod or in continuous darkness. The cell biomass was subcultured regularly into fresh medium at every three weeks.

iv. Batch kinetics of cell suspension cultures

a. Batch kinetics studies

To determine the specific growth rate, cells were harvested from liquid medium at an interval of two days, washed and dried. The pH and conductivity of the suspension cultures were monitored after every two days. Phosphate was estimated by the standard calibration curve made from dihydrogen sodium phosphate (NaH_2PO_4); to 0.5 ml of standard or sample solution, 4 ml of reagent [Acetone (CH_3COCH_3), Sulphuric acid (H_2SO_4)2.5 M and Ammonium molybdate tetrahydrate ($(NH_4)_2MoO_4.4H_2O$) 10 mM, mixed in the ratio of 2:1:1] was added. After mixing the solutions thoroughly, 0.4 ml of 1M citric acid was added and absorbance was taken at 355 nm. Similarly, for nitrate estimation, standard curve was made from 0.01N stock solution of Potassium nitrate (KNO_3), preserved in chloroform. After acidification of samples with hydrochloric acid, absorbance was recorded at 275 nm in a UV visible spectrophotometer. All measurements and results are average readings obtained from three flasks.

b. Agitation speed and cell viability

Effect of agitation speeds was evaluated on fresh and dry weight of cells and their viability, at the end of each passage. Callus cells weighing approximately 0.2 g were harvested at the end of growth period and re-inoculated in 50 ml of fresh medium of the same composition. The cultures were incubated in shaking conditions at 60, 120 and 240 rpm, under darkness, for a period of three weeks and their fresh and dry weights were recorded. The viability of cells under each condition was checked with 1% fluorescein diacetate (FDA) solution. FDA is a cell permeant dye. Within the cells, the molecule is cleaved by esterase activity to fluorescein which is unable to pass through the cell membrane of live cells while it leaches out from the dead cells. Hence, only the live, intact cells take up the stain and fluoresce green.

Results

Analysis of Callus Cultures

Establishing cell and tissue cultures of *Lantana* is a difficult task to accomplish because of the interference posed by phenolic compounds. The best treatment for callusing, in terms of number of explants showing callusing and the degree of callusing, was the combination of MS + 2,4-D (1 μM) + NAA (1 μM) + BAP (5 μM). On the responding medium, 100% of the explants callused, and the callus growth was profuse after the first subculture. On this medium, the leaf-disc explants first turned brown but after a week, bright-green, hard, compact calli started developing from the margins of the leaf-disc. These compact calli were dissected out and subcultured on the fresh medium of the same growth regulator composition. The rate and degree of callus proliferation increased with the subsequent subcultures, the nature of callus did not improve substantially. The callus was friable and soft but remained deep brown in nature. Until 10 th subculture, the cells in the callus were a mixture of green and brown cells. It took about 26 weeks of regular subculturing, at 4-week intervals, to obtain profusely growing fresh, friable, granulated and cream callus.

Callus culture of *Lantana camara* raised from leaf-disc explants

Analysis of Cell Suspension Cultures Raised in Shake Flasks

i.   Kinetics of cell growth and nutrient uptake

The specific growth rate (μ) of the suspended cells was found to be 0.1072 d^{-1}. It was observed that the cultures remained in the lag phase till the 2nd day. Biomass increased till the 10th day following which the stationary phase started. The pH of the medium underwent variation during different stages of culture. It was observed that after showing a slight decrease in the value, the pH dropped sharply between 4-6 days, which dropped further after 10 days resulting in poor growth and stationary phase of cells. The drop of pH may be attributed to preferential uptake of NH_4^+ ions which in turn resulted in decreased pH due to liberation of H^+ ions; pH tends to increase if NO_3^- is utilized faster than NH_4^+. Uptake of nitrate was observed at slower rate. It was present in the culture medium till the last days of cultivation (16th day). The concomitant synthesis of acid triterpenes in the medium was found to be growth associated and showed an increase with the increase in biomass. Conductivity was shown to have an inverse relationship with growth. Among the major inorganic nutrients, it was invariably observed that phosphate was almost completely consumed by the 10th day of culture. Its utilization was very fast in the initial days than in the later stages of growth. Hence, it may be concluded that complete utilization of phosphate and a marked drop in the pH of the culture medium resulted in the onset of stationary phase in cell suspension cultures.

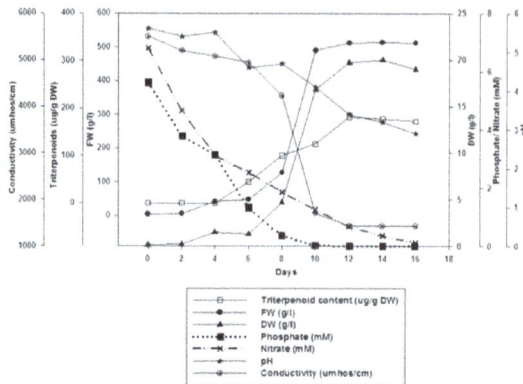

Kinetics of cell growth and nutrient uptake in cell suspension cultures of *L. camara* .

ii. Effect of agitation speed on cell survival and viability

Speed of agitation directly affected the growth and viability of cells in suspension cultures due to aeration and shearing effect. The maximum fresh weight and maximum viability was observed at 120 rpm. At lower agitation (60 rpm), the cells died due to aggregation and clumping; only the cells at the outermost layer of the aggregate were alive and fluorescent green when stained with fluorescein diacetate (FDA). At higher speed (240 rpm), the cells died due to rupturing.

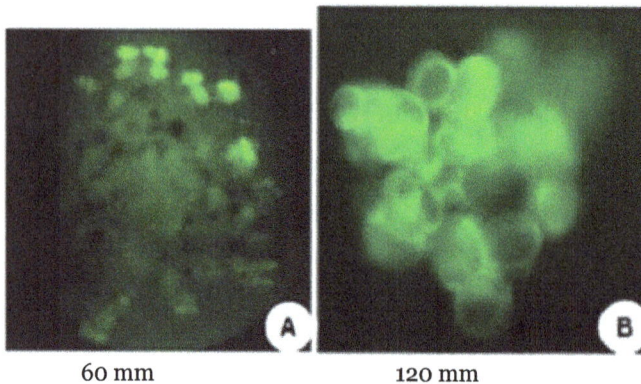

60 mm   120 mm

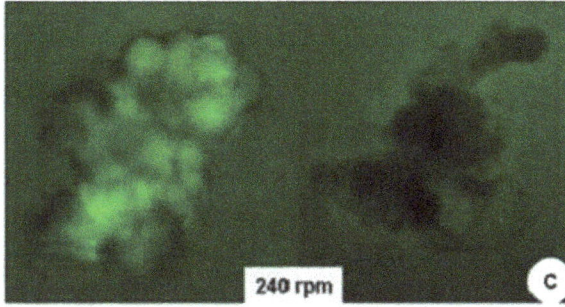

Cells stained with 1% fluorescein diacetate solution . A. cellular clump at 60 rpm, showing unstained dead cells in the centre of cell aggregate and live cells fluoresce green at the periphery , B. The cultures maintained at 120 rpm in the cell suspension, showing aggregates of live and healthy, fluorescent green stained cells , C. Same at 240 rpm, showing dead clumps (dark bodies) and sheared cells.

3D Cell Culturing by Magnetic Levitation

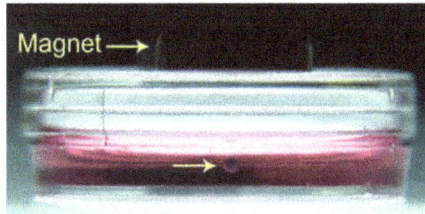

3D cell culture grown with magnetic levitation. Human Glioblastoma (hGBM) cells (indicated by lower arrow) treated with magnetic nanoparticles were held at the air-medium interface by the magnetic field created by the magnet attached to the top of the 1st tissue culture plate. This image was taken after 48 hr of culturing.

3D cell culture by the magnetic levitation method (MLM) is the application of growing 3D tissue by inducing cells treated with magnetic nanoparticle assemblies in spatially varying magnetic fields using neodymium magnetic drivers and promoting cell to cell interactions by levitating the cells up to the air/liquid interface of a standard petri dish. The magnetic nanoparticle assemblies consist of magnetic iron oxide nanoparticles, gold nanoparticles, and the polymer polylysine. 3D cell culturing is scalable, with the capability for culturing 500 cells to millions of cells or from single dish to high-throughput low volume systems. Once magnetized cultures are generated, they can also be use as the building block material, or the "ink", for the magnetic 3D bioprinting process.

Overview

Standard monolayer cell culturing on tissue culture plastic has notably improved our understanding of basic cell biology, but it does not replicate the complex 3D architecture of in vivo tissue, and it can significantly modify cell properties. This often compromises experiments in basic life science, leads to misleading drug-screening results on efficacy and toxicity, and produces cells that may lack the characteristics needed for developing tissue regeneration therapies.

The future of cell culturing for fundamental studies and biomedical applications lies in the creation of multicellular structure and organization in three-dimensions. Many schemes for 3D culturing are being developed or marketed, such as bio-reactors or protein-based gel environments.

A 3D cell culturing system known as the Bio-Assembler™ uses biocompatible polymer-basedreagents to deliver magnetic nanoparticles to individual cells so that an applied magnetic driver can levitate cells off the bottom of the cell culture dish and rapidly bring cells together near the air-liquid interface. This initiates cell-cell interactions in the absence of any artificial surface or matrix. Magnetic fields are designed to rapidly form 3D multicellular structures in as little as a few hours, including expression of extracellular matrix proteins. The morphology, protein expression, and response to exogenous agents of resulting tissue show great similarity to in vivo results.

History

3D cell culturing by magnetic levitation method (MLM) was developed from collaboration between scientists at Rice University and University of Texas MD Anderson Cancer Center in 2008. Since then, this technology has been licensed and commercialized by Nano 3D Biosciences.

The Magnetic Levitation Process

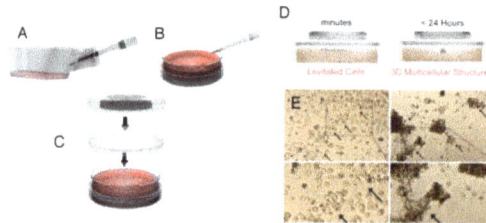

Above is a picture showing 3D cell culturing through magnetic levitation with the Bio-Assembler cell culturing system. (A) A magnetic iron oxide nanoparticle assembly known as Nanoshuttle is added and dispersed over cells and the mixture is incubated. (B) After incubation with Nanoshuttle, cells are detached and transferred to a petri dish. (C) A magnetic drive is then placed on top of a petri dish top. (D) The magnetic field causes cells to rise to the air–medium interface. (E) Human umbilical vein endothelial cells (HUVEC) levitated for 60 minutes (left images) and 4 hours (right images) (Scale bar, 50 μm). The onset of cell-cell interaction takes place as soon as cells levitate, and 3D structures start to form. At 1 hour, the cells are still relatively dispersed, but they are already showing some signs of stretching. Formation of 3D structures is visible after 4 hours of levitation (arrows).

Protein Expression

Protein expression in levitated cultures shows striking similarity to in vivo patterns. N-cadherin expression in levitated human glioblastoma cells was identical to the expression seen in human tumor xenografts grown in immunodeficient mice, while standard 2D culture showed much weaker expression that did not match xenograft distribution as shown in the picture below. The transmembrane protein N-cadherin is often used as an indicator of in-vivo-like tissue assembly in 3D culturing.

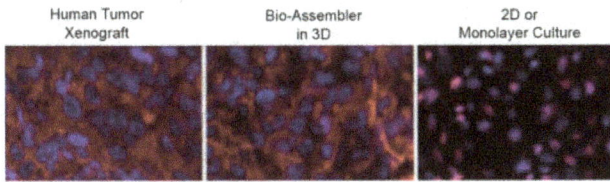

In the picture above, distribution of N-cadherin (red) and nuclei (blue) in human brain cancer mouse xenograft (left, human brain cancer cells grown in a mouse brain), brain cancer cells cultured by 3D magnetic levitation for 48 h. (middle), and cells cultured on a glass slide cover slip (2D, right). The 2D system shows N-cadherin in the cytoplasm and nucleus and notably absent from the membrane, while in the levitated culture and mouse, N-cadherin is clearly concentrated in the membrane, and also present in cytoplasm and cell junctions.

Applications

Co-culturing, Magnetic Manipulation, and Invasion Assays

One of the challenges in generating in vivo like cultures or tissue in vitro is the difficulty in co-culturing different cell types. Because of the ability of 3D cell culturing by magnetic levitation to bring cells together, co-culturing different cell types is possible. Co-culturing of different cell types can be achieved at the onset of levitation, by mixing different cell types in before levitation or by magnetically guiding 3D cultures in an invasion assay format.

The unique ability to manipulate cells and shape tissue magnetically offers new possibilities for controlled co-culturing and invasion assays. Co-culturing in a realistic tissue architecture is critical for accurately modeling in vivo conditions, such as for increasing the accuracy of cellular assays as shown in the figure below.

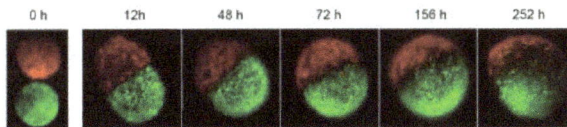

Shown in the picture above is an invasion assay of magnetically levitated multicellular spheroids. Fluorescence images of human glioblastoma (GBM) cells (green; GFP-expressing cells) and normal human astrocytes (NHA) (red; mCherry-labelled) cultured separately and then magnetically guided together (left, time 0). Invasion of GBM into NHAin 3D culture provides a powerful new assay for basic cancer biology and drug screening (right, 12h to 252h).

Vascular Simulation with Stem Cells

By facilitating assembly of different populations of cells using the MLM, consistent generation of organoids termed adipospheres capable of simulating the complex intercellular interactions of endogenous white adipose tissue (WAT) can be achieved.

Co-culturing 3T3-L1 pre-adipocytes in 3D with murine endothelial bEND.3 cells creates a vascular-like network assembly with concomitant lipogenesis in perivascular cells.

In addition to cell lines, WAT organogenesis can be simulated from primary cells.

Adipocyte-depleted stromal vascular fraction (SVF) containing adipose stromal cells (ASC), endothelial cells, and infiltrating leukocyte derived from mouse white adipose tissue (WAT) were cultured in 3D. This revealed organoids striking in hierarchical organization with distinct capsule and internal large vessel-like structures lined with endothelial cells, as well as perivascular localization of ASC.

Upon adipogenesis induction of either 3T3-L1 adipospheres or adipospheres derived from SVF, the cells efficiently formed large lipid droplets typical of white adipocytes in vivo, whereas only smaller lipid droplet formation is achievable in 2D. This indicates intercellular signaling that better recapitulates WAT organogenesis.

This MLM for 3D co-culturing creates adipospheres appropriate for WAT modeling ex vivo and provides a new platform for functional screens to identify molecules bioactive toward individual adipose cell populations. It can also be adopted for WAT transplantation applications and aid other approaches to WAT-based cell therapy.

Organized co-culturing to Create in Vivo-like Tissue

Using the MagPen™ (a Nano3D Biosciences, Inc. product), organized 3D co-cultures similar to native tissue architecture can be rapidly created. Endothelial cells (PEC), smooth muscle cells (SMC), fibroblasts (PF), and epithelial cells (EpiC) cultured with the Bio-Assembler™ can be sequentially layered in a drag-and-drop manner to create bronchioles that maintain phenotype and induce extracellular matrix formation.

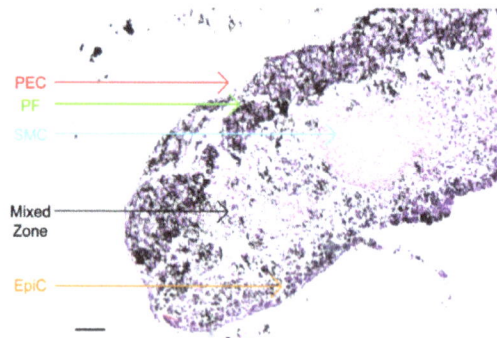

Organized bronchiole created with MagPen™ and Bio-Assembler™. Endothelial cells (PEC), smooth muscle cells (SMC), fibroblasts (PF), and epithelial cells (EpiC) can be sequentially layered. Scale bar: 100um.

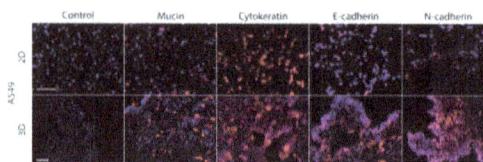

IHC reveals epithelial markers on A549 after culturing in 3D with the MLM

Cell Types Cultured

Listed below are the cell types (primary and cell lines) that have been successfully cultured by the magnetic levitation method. The second table is the same but with images included. More images are available at Nano3D Biosciences, Inc.

Cells	Cell line	Organism	Organ tissue
Murine endothelial	Cell line	Mouse	Vessel
Murine adipocyte	Cell line	Mouse	Adipose
Rattus norvegicus hepatoma	Cell line	Rat	Liver
Pulmonary fibroblasts (HPF)	Primary	Human	Lung
Pulmonary endothelial (HPMEC)	Primary	Human	Lung
Small airway epithelial (HSAEpiC)	Primary	Human	Lung
Bronchial epithelial	Primary	Human	Lung
Human alveolar adenocarcinoma	A549	Human	Lung
Type II alveolar	Primary	Human	Lung
Tracheal smooth muscle (HTSMC)	Primary	Human	Lung
Mesenchymal stem cells (HMSC)	Primary	Human	Bone marrow
Bone marrow endothelial cells (HBMEC)	Primary	Human	Bone marrow
Dental pulp stem cells	Primary	Human	
Human umbilical vein endothelial cells (HUVEC)	Primary	Human	
Murine chondrocytes	Primary	Mouse	Bone
Murine adipose tissue	Primary	Mouse	
Heart valve endothelial	Primary	Porcine	
Pre-adipocytes fibroblasts	3T3	Mouse	
Neural stem cells	C17.2	Mouse	Brain
Human embryonic kidney cells	HEK293	Human	Kidney
Melanoma	B16	Mouse	Skin
Astrocytes	NHA	Human	Brain
Glioblastomas	LN229	Human	Brain
T-cells and antigen presenting cells		Human	
Mammary epithelial	MCF10A	Human	Breast
Breast cancer	MDA231	Human	Breast
Osteosarcoma	MG63	Human	Bone

Protoplast Isolation and Regeneration

The term protoplast was introduced by Hanstein in 1880. It refers to the cellular content excluding cell wall or can also be called as naked plant cell. It is described as living matter enclosed by a plant cell membrane. Protoplast isolation for the first time was carried out by Klercker in 1892 using mechanical method on the plasmolysed cells. The application of protoplast technology for the improvement of plants offers fascinating option to complement conventional breeding programs. The ability of isolated protoplasts to undergo fusion and take up macromolecules and cell organelles offers many possibilities in genetic engineering and crop improvement (Bhojwani et al . 1977). The experiments involving protoplasts consist of three stages –

i.   protoplast isolation

ii. protoplast fusion (leading to gene uptake)

iii. development of regenerated fertile plants from the fusion product (Hybrid).

Depending upon the species and culture conditions, the protoplasts may have the potential to:

i.   regenerate a cell wall

ii. dedifferentiate to form callus

iii. divide mitotically and proliferate clonally

iv   redifferentiate into shoots, roots or embryos and produce a complete plantlet.

However, to fully explore the potentials for protoplast-technology, efficient and reproducible methods for protoplast isolation and purification must first be established. Since leaf tissue is a readily accessible source of genetically uniform cells, it is often desirable to use mesophyll protoplasts in somatic hybridization studies, but, leaf tissues, in general, do not yield large number of protoplasts owing to the difficulty in removing the lower epidermis (Chaturvedi 2003). An alternative, therefore, is the cultured cell material where protoplasts can show greater potential to divide (Bhojwani and Razdan 1996).

Protoplast Isolation

Protoplast isolation may be carried out by Mechanical disruption method or enzymatic method. Out of these two methods, enzymatic method is preferred as it provides better protoplast yield with low tissue damage while mechanical method causes maximum tissue chopping with lower protoplast yields. Both of these methods are described below:

i.   Mechanical method

Klercker in 1892 pioneered the isolation of protoplasts by mechanical methods. In this method, the cells were kept in suitable plasmolyticum, for example CPW containing 13% w/v mannitol. Once the plasmolysis is complete, while remaining in the osmoticum, the leaf lamina would be cut with a sharp-edged knife. In this process some of the plasmolyzed cells were cut only through the cell wall, releasing intact protoplasts while some of the protoplasts may be damaged inside many cells. Protoplasts that were trapped in a cell and only the corner had been cut off could be encouraged to come out by reducing the osmolarity slightly to force the protoplasts swell to force their way out of the cut surface. The released protoplasts then have to be separated from damaged ones and cell debris.

Disadvantages

- Lower protoplast yield.

- Labour intensive method.

- Protoplast obtained has low viability.

- Method is applicable only to vacuolated cells.

ii. Enzymatic method

In 1960, E.C. Cocking demonstrated the possibility of enzymatic isolation of a large number of protoplasts from cells of higher plants. This method involves leaf sterilization followed by peeling of the lower epidermis to release cells which are plasmolyzed and added to enzyme mixture followed by harvest of protoplast as shown in. Either of the procedures for enzymatic isolation can be used: sequential enzymatic hydrolysis or mixed enzymatic hydrolysis.

In the sequential isolation, firstly, cells are separated by the use of a maceration enzyme – a pectin hydrolyzing enzyme such as, macerozyme or Pectolyase. Once the cells are separated, they are washed in CPW solution free of enzymes but containing plasmolyticum by gentle centrifugation (100g). The pellet is retained and resuspended in the second enzyme like, cellulases and hemicellulases, used to hydrolyse the remaining cell was component. Once the protoplasts are released they are washed with CPW to remove the debris.

In the mixed enzymatic approach, Plant tissues are plasmolyzed in the presence of a mixture of pectinases and cellulases, thus, inducing simultaneous separation of cells and degradation of their walls to release the protoplasts directly in a single step.

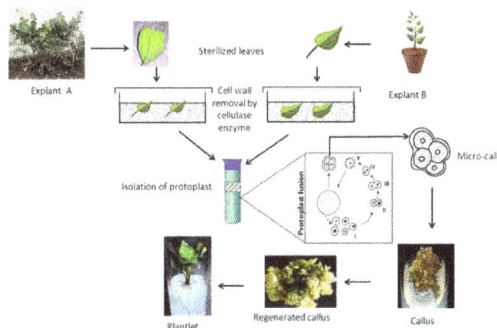

Steps involved in protoplast isolation, fusion and regeneration

Conditions required for enzyme activity :

- Enzymes are pH and temperature dependent, thus, for enzymatic release of protoplast an enzyme showing activity at pH range 4.7-6.0 and temperature range of 25-30°C is used

- Duration of enzyme pretreatment and condition of light presence required for incubation may also be determined.

- Enzyme mixture used should essentially consist of cellulose, hemicellulase and pectinase which facilitate the degradation of cellulose, hemicelluloses and pectin, respectively.

- The concentration of sugar alcohols used as osmoticum (mannitol) must be empirically defined.

Factors Affecting Yield and Viability of Protoplasts

i. *Source of material* : Leaves were the most convenient source of the plant protoplasts because it allows the isolation of a large number of relatively uniform cells without killing the plants. Moreover the mesophyll cells are loosely arranged, the enzymes have an easy access to the cell wall. The parent plant age and the conditions in which it is growing have profound effect on the yield of protoplast. Due to the difficulty in isolating culturable protoplast from leaf cells of cereals and some other species their cultured cells can be used as a source material. The yield of protoplasts depends upon the growth rate and growth phase of the cells. Generally embryogenic suspension cultures are used to obtain totipotent protoplasts .

ii. *Pre-enzyme treatments* : To facilitate the penetration of enzyme solution into the intercellular spaces of leaf, which is essential for effective digestion, various methods are followed. The most commonly used method is to peel the lower epidermis and float the stripped pieces of leaf on the enzyme solution in a manner that the peeled surface is in a contact with the solution. Most of the time it is not convenient to peel the epidermis, in such cases cutting the leaf or tissue into small strips (1- 2 mm wide) has been found useful. When combined with vacuum infiltration the latter approach has proved very effective. Brushing of leaves with a soft brush or with the cutting edge of a scalpel may also improve the enzymatic action. Large calli are chopped into pieces and can be transferred to enzyme mixture. Agitation of incubation mixture during enzyme treatment improves protoplast yield from cultured cells.

iii. *Enzyme treatment* : The release of protoplast is very much dependent on the nature and concentration of the enzymes used. The two major enzymes required for the isolation of protoplast are cellulase and pectinase. The cellulase is required to digest the cellulosic cell walls and the pectinase mainly degrades the middle lamella. Some of the tissues may require other enzymes like, hemicellulase, driselase, macerozyme and pectolyase. The activity of enzyme is pH dependent. The pH of the enzyme solution is adjusted somewhere between 4.7 to 6.0.

Choice of Plasmolyticum

The two most commonly used compounds are the sugar alcohols - mannitol and sorbitol. Of these, mannitol is the most preferred since it is not metabolized by the plant cells. Once the protoplast divides and regenerates the cell wall, no more osmoticum is required. It is, therefore, should be removed gradually from the medium otherwise cell division stops. To slowly remove the osmoticum from the medium, the protoplast can be isolated in a high osmoticum mixture consisting of both mannitol and sucrose, the sucrose will be metabolized by the dividing protoplasts and thus, will reduce the osmolarity of the medium. Normally, mannitol is used at concentration range of 11-13%.

A solution into which the osmoticum is often, but not always, added is called CPW salts mix or CPW for short. This has been observed much more beneficial than using distilled water as a solvent in obtaining high yields of viable protoplasts:

Table: Salt mix of protoplast washing media solution (Cocking, Peberdy and White – CPW)

KH_2PO_4	27.2 mg
KNO_3	101 mg
$CaCl_2.2H_2O$	1480 mg
$MgSO_4.7H_2O$	246 mg
KI	0.16 mg
$CuSO_4.5H_2O$	0.025 mg
Made to 1 litre with water : pH 5.8	

Although CPW is most widely used solution into which osmoticum or enzymes are added, some times culture medium used to grow cells or plants can also be utilized for protoplast isolation at one tenth concentration. Low concentration of culture medium is much more advantageous when compared with CPW.

Protoplast Purification

Enzyme treatment results in suspension of protoplast, undigested tissues and cellular debris. This suspension is passed through a metal sieve or a nylon mesh (50-100 µm) in order to remove undigested cellular clumps. The filtered protoplast-enzyme solution is mixed with a suitable volume of osmoticum, solution is centrifuged to pellet the protoplasts, pellet of protoplast is resuspended in osmoticum of similar concentration as used in enzyme mixture. The protoplast band is sucked in Pasteur pipette and is put into other centrifuge and finally suspended in culture medium at particular density; this is explained by the figure.

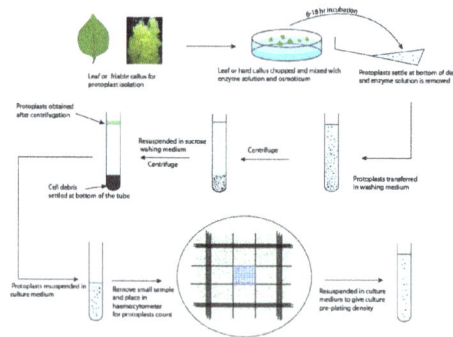

schematic diagram of protoplast purification

Protoplast viability: The isolated protoplast must have a spherical shape when observed by a light microscope, protoplast can be stained using following stain:

- *Fluorescein diacetate staining method:* FDA accumulates inside the plasma-lemma of viable protoplasts. Live protoplasts contain esterases which cleave FDA to release fluorescein which fluoresces yellowish-green using fluorescence microscopy within 5 min. FDA dissociates from membrane after about 15 min. It is used at a concentration of 0.01% dissolved in acetone.

- *Calcofluor White staining:* This staining method assures protoplast viability by detecting onset of cell wall formation. Calcofluor binds to beta linked gluco-sides in newly synthesized cell wall which can be observed as a fluorescent ring around the membrane. Optimum staining is achieved when 0.1 ml of protoplast is mixed with 5.0 µl of 0.1% w/v solution of CFW.

- *Protoplast viability* can also be detected by monitoring oxygen uptake of cells by oxygen electrode, which shows respiration.

- *Variation of protoplast size* with changing osmotic concentration also enables viability of protoplast.

Protoplast Culture

Protoplasts Culture Techniques

The culture requirements and the culture methods are same for both protoplasts and single cells. The main difference is the requirement of suitable osmoticum for pro-toplasts until they regenerate a strong wall. Isolated protoplasts are either cultured in liquid or semisolid agar or agarose media plates, sometimes the protoplast is first grown in liquid media and then transferred into the agar media plates. The following techniques have been adopted in order to maintain number of protoplast population between minimum and maximum effective densities after plating up:

i. *Liquid method* : This method is preferred in earlier stages of culture as it provides (a) easier dilution and transfer, (b) the osmotic pressure of liquid media can be effectively reduced after a few days of culture (c) the cell density can be reduced or cells of special interest can be isolated easily. In Liquid medium, the protoplast suspension is plated as a thin layer in petriplates, incubated as static culture in flasks or distributed in 50-100 µl drops in petriplates and stored in a humidifier chamber.

ii. *Embedded in Agar/Agarose* : Agarose is a preferred choice in place of agar and this has improved the culture response. This method of agar culture keeps protoplast in fixed position, thus, prevents it from forming clumps. Immobilized protoplasts give rise to clones which can then be transferred to other media. In practice, the protoplasts suspended in molten (40°C) agarose medium (1.2% w/v agarose) are dispensed (4ml) into small (3.5-5cm diameter) plates and allowed to solidify. The agarose layer is then cut into 4 equal sized blocks and transferred to larger dishes (9 cm diameter) containing liquid medium of otherwise the same composition. Alternatively, protoplasts in molten agarose medium are dispensed as droplets (50-100 µl) on the bottom of petri plates and after solidification the droplets are submerged in the same liquid medium.

iii. *Feeder layer* : In order to culture protoplast at low density, a feeder layer technique is adopted. A feeder layer of X-ray irradiated non-dividing but metabolically active protoplasts after washing are plated in soft agar medium. Non-irradiated protoplasts of low density are plated over this feeder layer. The protoplasts of the same species or different species can be used as a feeder layer.

iv. *Co-culturing* : This method involves co-culture of protoplasts from two different species to promote their growth or that of the hybrid cells. Metabolically active and dividing protoplasts of two types - slow and fast growing are cultured together, the fast growing protoplast provide other species with diffusible chemicals and growth factors which helps in cell wall formation and cell division. The co-culture methods is generally used where calli arising from two types of protoplasts can be morphologically distinguished. For example, protoplasts isolated from albino plants and green plants are easily distinguishable based on color where albino protoplast will develop non green colonies.

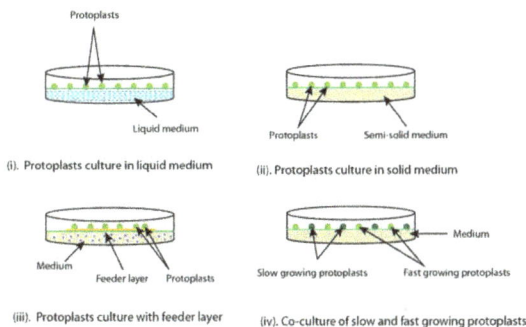

(i). Protoplasts culture in liquid medium

(ii). Protoplasts culture in solid medium

(iii). Protoplasts culture with feeder layer

(iv). Co-culture of slow and fast growing protoplasts

protoplast culture techniques

Culture Medium

The nutritional requirement of protoplast is almost similar to that of the cultured plant cells. Mostly the salts of MS (Murashige and Skoog, 1962) and B_5 (Gamborg et al. 1968) media and their modifications have been used. Ammonium salts have been found detrimental to protoplasts survival of many species, and media have been devised that either have a reduced concentration of ammonium or lack it. Concentration of zinc is reduced while the concentration of calcium is increased as it enhances the membrane stability. Osmolarity is maintained by addition of sorbitol, mannitol, glucose or sucrose and mannitol being widely used osmoticum as it is not used by the dividing cells, thus, maintains the osmolarity of the medium.

Glucose is preferred carbon source as sucrose do not satisfy protoplast culture. One or two amino acids are added at low concentration. Growth regulators are required essentially in protoplast culture generally high concentration of auxins (NAA, 2,4-D) along with lower concentration of cytokinins (BAP, Zeatin) is used.

Environmental conditions: High light intensity inhibits growth of protoplast hence initially protoplast is grown in dim light for few days and then transferred to light of about 2000-5000 lux. However, better results are obtained when cultured in darkness.

Plating Density

Like cell cultures, the initial plating density of protoplasts has profound effect on plating efficiency. Protoplasts are cultured at a density of 1×10^4 to 1×10^5 protoplasts ml^{-1} of the medium. At high density the cell colonies arising from individual protoplasts tend to grow into each other resulting into chimera tissue if the the protoplast population is genetically heterogeneous. Cloning of individuals cells, which is desirable in somatic hybridization and mutagenic studies, can be achieved if protoplasts or cells derived from them can be cultured at a low density.

Protoplast Development and Regeneration

Protoplast starts to regenerate a cell wall within few days (2-4 days) of culture and during this process, protoplasts lose their characteristic spherical shape which has been taken as an indication of new wall regeneration. Cell wall regeneration can be confirmed by Calcofluor White staining method. There is direct relationship between wall formation and cell division. Protoplasts which are not able to regenerate a proper wall fail to undergo normal mitosis. Protoplasts with a poorly developed wall often show budding and may enlarge several times their original volume. They may become multinucleate because karyokinesis is not accompanied by cytokinesis. Among other reasons, inadequate washing of the protoplasts prior to culture leads to these abnormalities.

And completes process when provided with suitable condition of light, pH and temperature newly synthesized protoplast can be visualized by staining. Once the cell wall formation is completed, cells undergo division resulting in increase size of cells. After an interval of 3 weeks, small cell colonies appear, these colonies are transferred to an osmotic-free callus induction medium. This is followed by introduction into organogenic or embryogenic medium leading to plantlet development.

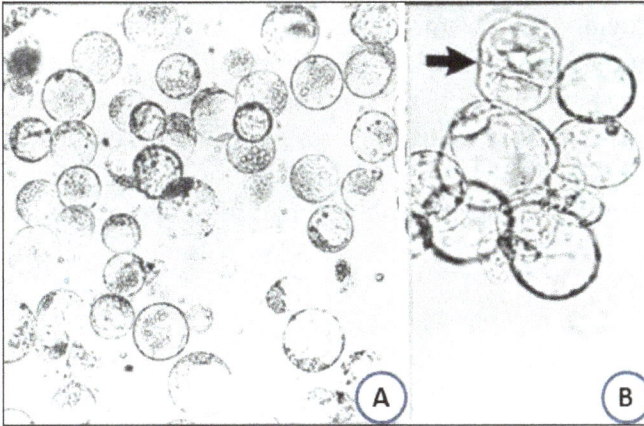

Protoplast isolation and cell wall regeneration. A. Isolated protoplast showing spherical structure; B. Wall is regenerated around the protoplast and one of the protoplasts showing cell division (arrow marked)

Somatic Hybridization and Cybridization

Sexual hybridization since time immemorial has been used as a method for crop improvement but it has its own limitations as it can only be used within members of same species or closely related wild species. Thus, this limits the use of sexual hybridization as a means of producing better varieties. Development of viable cell hybrids by somatic hybridization, therefore, has been considered as an alternative approach for the production of superior hybrids overcoming the species barrier. The technique can facilitate breeding and gene transfer, bypassing problems associated with conventional sexual crossing such as, interspecific, intergeneric incompatibility. This technique of hybrid production via protoplast fusion allows combining somatic cells (whole or partial) from different cultivars, species or genera resulting in novel genetic combinations including symmetric somatic hybrids, asymmetric somatic hybrids or somatic cybrids.

The most common target using somatic hybridization is the gene of symmetric hybrids that contain the complete nuclear genomes along with cytoplasmic organelles of both parents. This is unlike sexual reproduction in which organelle genomes are generally contributed by the maternal parent. On the other hand, somatic cybridization is the process of combining the nuclear genome of one parent with the mitochondrial and/

or chloroplast genome of a second parent. Cybrids can be produced by donor-recipient method or by cytoplast-protoplast fusion. Incomplete asymmetric somatic hybridization also provides opportunities for transfer of fragments of the nuclear genome, including one or more intact chromosomes from one parent (donor) into the intact genome of a second parent (recipient).

These methods provide genetic manipulation of plants overcoming hurdle of sexual incompatibility, thereby, serving as a method of bringing together beneficial traits from taxonomically distinct species which cannot be achieved by sexual crosses. Several parameters such as, source tissue, culture medium and environmental factors influence ability of a protoplast derived hybrid cells to develop into a fertile plant. The general steps involved in somatic hybridization and cybridization methods are elaborated in Figur.

Steps Involved in Somatic Hybridization

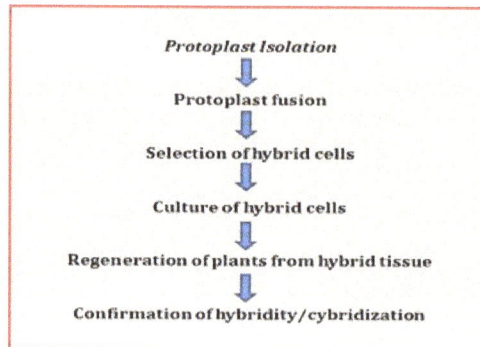

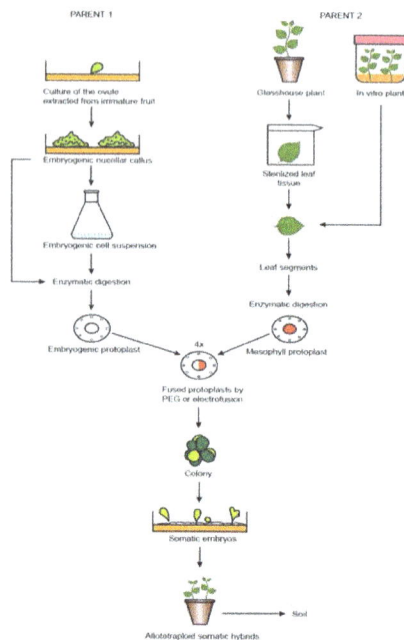

Schematic view of symmetric protoplast fusion producing somatic hybrids.

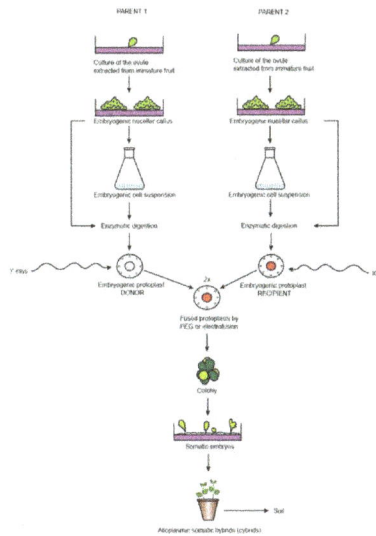

Schematic view of asymmetric protoplast fusion using donor-recipient method resulting into creation of alloplasmic somatic hybrid or cybrids.

Protoplast Fusion

Protoplast fusion could be spontaneous during isolation of protoplast or it can be induced by mechanical, chemical and physical means. During spontaneous process, the adjacent protoplasts fuse together as a result of enzymatic degradation of cell walls forming homokaryons or homokaryocytes, each with two to several nuclei. The occurrence of multinucleate fusion bodies is more frequent when the protoplasts are prepared from actively dividing callus cells or suspension cultures. Since the somatic hybridization or cybridization require fusion of protoplasts of different origin, the spontaneous fusion has no value. To achieve induced fusion, a suitable chemical agent (fusogen) like, $NaNO_3$, high Ca^{2+}, polyethylene glycol (PEG), or electric stimulus is needed.

i.   Fusion by means of $NaNO_3$: It was first demonstrated by Kuster in 1909 that the hypotonic solution of $NaNO_3$ induces fusion of isolated protoplast forming heterokaryon (hybrid). This method was fully described by Evans and Cocking (1975), however this method has a limitation of generating few no of hybrids, especially when highly vacuolated mesophyll protoplasts are involved.

ii   High pH and Ca^{++} treatment: This technique lead to the development of intra- and interspecific hybrids. It was demonstrated by Keller and Melcher in 1973. The isolated protoplasts from two plant species are incubated in 0.4 M mannitol solution containing high Ca^{++}(50 mM $CaCl_2.2H_2O$) with highly alkaline pH of 10.5 at 37°C for about 30 min. Aggregation of protoplasts takes place at once and fusion occurs within 10 min.

iii Polyethylene glycol treatment: Polyethylene glycol (PEG) is the most popularly known fusogen due to ability of forming high frequency, binucleate heterokaryons with low cytotoxicity. With PEG the aggregation occurred mostly between two to three protoplasts unlike Ca^{++} induced fusion which involves large clump formation. The freshly isolated protoplasts from two selected parents are mixed in appropriate proportions and treated with 15-45% PEG (1500-6000MW) solution for 15-30 min followed by gradual washing of the protoplasts to remove PEG. Protoplast fusion occurs during washing. The washing medium may be alkaline (pH 9-10) and contain a high Ca^{++} ion concentration (50 mM). This combined approach of PEG and Ca^{++} is much more efficient than the either of the treatment alone. PEG is negatively charged and may bind to cation like Ca^{++}, which in turn, may bind to the negatively charged molecules present in plasma lemma, they can also bind to cationic molecules of plasma membrane. During the washing process, PEG molecules may pull out the plasma lemma components bound to them. This would disturb plamalemma organization and may lead to the fusion of protoplasts located close to each other. The technique is nonselective thus, induce fusion between any two or more protoplasts.

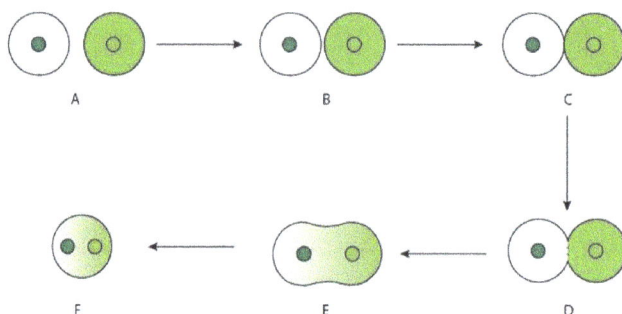

Sequential stages in protoplast fusion. (A) two separate protoplasts, (B) agglutination of two protoplasts, (C and D) Membrane fusion at localized site, and (E and F) development of spherical heterokaryon.

iv. Electrofusion: The chemical fusion of plant protoplast has many disadvantages – (1) The fusogen are toxic to some cell systems, (2) it produces random, multiple cell aggregates, and (3) must be removed before culture. Compare to this, electrofusion is rapid, simple, synchronous and more easily controlled. Moreover, the somatic hybrids produced by this method show much higher fertility than those produced by PEG-induced fusion.

Zimmermann and Scheurich (1981) demonstrated that batches of protoplasts could be fused by electric fields by devising a protocol which is now widely used. This protocol involves a two-step process. First, the protoplasts are introduced into a small fusion chamber containing parallel wires or plates which serve as electrodes. Second, a low-voltage and rapidly oscillating AC field is applied, which causes protoplasts to become aligned into chains of cells between electrodes. This creates complete cell-to-cell contact within a few minutes. Once alignment is complete, the fusion is induced by application of a brief spell of high-voltage DC pulses (0.125-1 kVcm^{-1}). A high voltage DC pulses induces a reversible breakdown of the plasma membrane at the site of cell contact, leading to fusion and consequent membrane reorganization. The entire process can be completed within 15 min.

Selection of Fusion Products

The somatic hybridization by electrofusion of protoplasts allow one-to-one fusion of desired pairs of protoplasts and, therefore, it is easy to know the fate of fusion products. However, protoplast suspension recovered after chemical treatments (fusogen) consists of the following cell types:

i.   unfused protoplasts of the two species/strains

ii. products of fusion between two or more protoplasts of the same species (homokaryons), and

iii. hybrid protoplasts produced by fusion between one (or more) protoplasts of each of the two species (heterokaryons)

The heterokaryons which are the potential source of future hybrids constitute of a very small (0.5-10%) proportion of the mixture. Therefore, an effective strategy has to be employed for their identification and isolation. Various protocols have been proposed and practiced for the effective selection of hybrids, including morphological basis, complementation of biochemical and genetic traits of the fusing partners, and manual or electronic sorting of heterokaryons/hybrid cells.

Morpho-physiological basis: The whole mixture of the protoplasts are cultured after fusion treatment and the resulting calli or regenerants are screened for their hybrid characteristics. Occasionally the hybrid calli outgrow the parental cell colonies and are identified by their intermediate morphology, i.e. green with purple coloured cells. However, the process is labour intensive and requires glasshouse facilities. It is limited to certain combinations showing differences in their regeneration potential under specific culture conditions.

Complementation: In this case complementation or genetic or metabolic deficiencies of the two fusion partners are utilized to select the hybrid component. When protoplasts of two parents, (one parent bearing cytoplasmic albino trait and the other parent bearing green trait) each parent carrying a non-allelic genetic or metabolic defect are fused, it reconstitutes a viable hybrid cell, of wild type in which both defects are mutually abolished by complementation, and the hybrid cells are able to grow on minimal medium non-permissive to the growth of the parental cells bearing green trait. Later, the calli of hybrid nature could be easily distinguished from the parental type tissue (albino trait) by their green color. The complementation selection can also be applied to dominant characters, such as dominant resistance to antibiotics, herbicides or amino acid analogues.

Isolation of heterokaryons or hybrid cells: The manual or electronic isolation of heterokaryons or hybrid cells is the most reliable method. Manual isolation requires that the two parental type protoplasts have distinct morphological markers and are easily distinguishable. For example, green vacuolated, mesophyll protoplasts from one parent and richly cytoplasmic, non green protoplasts from cultured cells of another parent. The dual fluorescence method also helps easy identification of fusion products. In this case, the protoplast labeled green by treatment with fluorescein diacetate (FDA, 1-20 mgl^{-1}) are fused with protoplasts emitting a red fluorescence, either from chlorophyll autofluorescence or from exogenously applied rhodamine isothiocyanate (10-20 mgl^{-1}). The labeling can be achieved by adding the compound into the enzyme mixture. This can be applied even for morphologically indistinguishable protoplasts from two parents.

Verification and Characterization of Somatic Hybrids

As no system is foolproof and they have their own advantages and disadvantages. Therefore, even after selecting the desired hybrids/cybrids following protoplast fusion, it is required to carry out one or more tests to compare the parent protoplast lines with the putative hybrids. Some of the techniques that can be tried are:

Morphology: Somatic hybrids in most of the cases show characters intermediate between the two parents such as, shape of leaves, pigmentation of corolla, plant height, root morphology and other vegetative and floral characters. The method is not much accurate as tissue culture conditions may also alter some morphological characters or the hybrid may show entirely new traits not shown by any of the parents.

Isozyme analysis: Multiple molecular forms of same enzyme which catalyses similar or identical reactions are known as isozymes. Electrophoresis is performed to study banding pattern as a check for hybridity. If the two parents exhibit different band patterns for a specific isozyme the putative hybrid can be easily verified. The isozymes commonly used for hybrid identification include, acid phosphatase, esterase, peroxidase.

Cytological analysis: Chromosome counting of the hybrid is an easier and reliable method to ensure hybridity as it also provides the information of ploidy level. Cytologically the chromosome count of the hybrid should be sum of number of chromosomes from both the parents. Besides number of chromosomes, the size and structure of chromosomes can also be monitored. However, the approach is not applicable to all species, particularly where fusion involves closely related species or where the chromosomes are very small. Moreover, sometimes the somaclonal variations may also give rise to different chromosome number.

Molecular analysis: Specific restriction pattern of nuclear, mitochondrial and chloroplast DNA characterizes the plastomes of hybrids and cybrids. Molecular markers such as RFLP, RAPD, ISSR can be employed to detect variation and similarity in banding pattern of fused protoplasts to verify hybrid and cybrid.

Cybrids or Cytoplasmic Hybrids:

Sexual hybridization involves fusion of the nuclear genes of both the parents but somatic hybrids involves even cytoplasm from both the parental species in hybrid obtained by protoplast fusion. However, in another case somatic hybrids containing nuclear genome of one parent but cytoplasm from both the parents, are termed as cybrids. The approach is time consuming and require several years of crossing plants provides an opportunity to study interparental mitochondrial, chloroplast fusion giving rise to plants with novel genomes.

Methods to produce cybrids: They are produced in variable frequencies in normal protoplast fusion experiments due to one of the following methods:

1. Fusion of normal protoplast with an enucleated protoplast. The enucleated protoplast can be produced by high speed centrifugation (20,000-40,000xg) for 60 min with 5-50% percoll.

2. Fusion between a normal protoplast and another protoplast with a non-viable nucleus or suppressed nucleus.

3. Elimination of one of the nuclei after heterokaryons formation.

4. Selective elimination of chromosomes at a later stage.

5. Irradiating (with X-rays or gamma rays) the protoplasts of one species prior to fusion in order to inactivate their nuclei.

6. By preparing enucleate protoplasts (cytoplasts) of one species and fusing them with normal protoplasts of the other species.

Cybrids provide the following unique opportunities: (i) transfer of plasmogenes of one species into the nuclear background of another species in a single generation, and even in (ii) sexually incompatible combinations, (iii) recovery of recombinants between the parental mitochondrial or chloroplast DNAs (genomes), and (iv) production of a wide variety of combinations of the parental and recombinant chloroplasts with the parental or recombinant mitochondria.

Applications of Somatic Hybridization

1. Novel interspecific and intergeneric crosses which are difficult to produce by conventional methods can be easily obtained.

2. Important characters, such as resistance to diseases, ability to undergo abiotic stress and other quality characters, can be obtained in hybrid plant by the fusion of protoplasts of plant bearing particular character to the other plant which may be susceptible to diseases.

3. Protoplasts of sexually sterile haploid, triploid, aneuploid plants can be fused to obtain fertile diploids and polyploids.

4. Studying cytoplasmic genes may be helpful to carry out plant breeding.

5. Most of the agronomically important traits, such as cytoplasmic male sterility, antibiotic resistance and herbicide resistance, are cytoplasmically encoded, hence can be easily transferred to other plant.

6. Plants in juvenile stage can also be hybridized by means of somatic hybridization.

7. Somatic hybridization can be used as a method for the production of autotetraploids.

Limitations of Somatic Hybridization

1. Application of protoplast methodology requires efficient plant regeneration system from isolated protoplasts. Protoplasts from two species can be fused, however, production of somatic hybrids is not easy.

2. Lack of a proper selection method for fused products (hybrids) poses a problem.

3. The end product of somatic hybridization are often unbalanced (sterile, misformed and unstable)

4. Somatic hybridization of two diploids leads to formation of amphidiploids which is unfavorable.

5. It is not sure for a character to completely express after somatic hybridization.

6. The regeneration products of somatic hybridization are often variable due to somaclonal variation, chromosome elimination, organelle segregation.

7. All diverse intergeneric somatic hybrids are sterile and, therefore, have limited chances of development of new varieties.

8. To transfer useful genes from wild species to cultivated crop, it is necessary to achieve intergeneric recombination or chromosome substitution between parental genomes.

Cryopreservation

In recent years, with the enormous increase in the population, pressure on the forest and the land resources have increased. This results in depletion of population of medicinal and aromatic plant species. Even some of the plant species are at the verge of vanishing from the forest. The list of endangered species is growing day by day. The conventional methods of germplasm preservation are prone to possible catastrophic losses because of:

i.   Attack by pathogen and pests.

ii. Climatic disorders

iii. Natural disasters and

iv. Political and economic causes

In addition, the seeds of many important medicinal plants lose their viability in a short time under conventional storage system.

The conservation of germplasm can be done by two methods:

1. In-situ preservation: Preservation of the germplasm in their natural environment by establishing biospheres, national parks etc.

2. Ex-situ preservation: In the form of seeds or by *in vitro* cultures.

Seeds form the most common material to conserve plant germplasm, however, the method has the following disadvantages:

i.   Discrete clones cannot be maintained in the form of seeds.

ii. Some plants do not produce fertile seeds.

iii. Loss of seed viability.

iv. Seed destruction by pests, etc.

v.   Poor germination rate.

vi. This is useful for seed propagating plants and is not applicable to vegetatively propagated crops, like potato, ginger etc.

In vitro preservation by tissue culture has several advantages over seed preservation:

i.   Large amount of materials can be stored in a small area.

ii. The material could serve as an excellent form of nucleus stock to propagate large number of plants rapidly, when required.

iii. Under special storage conditions the plants do not require frequent splitting and pruning.

iv. Being free from known viruses and pathogens, the clonal plant material could be sent from country to country, thus, minimizing the obstructions imposed by quarantine systems on the movement of live plants across national boundaries.

v. Protection from natural hazards.

vi. The plants are not exposed to the threat of changing government policies and urban development.

There are few disadvantages of in vitro system to be used for conservation of plant material:

i. It is a costly process.

ii. In cultures, plants can be maintained by serial subcultures at frequent intervals for virtually unlimited periods. However, the storage of germplasm by serial subcultures risks the loss of plant material by microbial contamination due to human error and also, is uneconomical. Moreover, in long-term callus and suspension cultures, the regeneration potential, biosynthetic properties and genetic make-up of the cells suffer. The basic requirement of a plant tissue culture method is the preservation of genetic resources, therefore, is to reduce the frequency of subcultures to a bare minimum.

Cryopreservation

Cryopreservation means preservation in the frozen state. The principle involved in cryopreservation is to bring the plant cell and tissue cultures to zero metabolism or non-dividing state by mean of storage of germplasm at a very low temperatures, (i) Over solid CO_2 (-79°C), (ii) Deep freezers (-80°C), (iii) in vapor phase nitrogen (-150°C), (iv) in liquid nitrogen (-196°C).

Among these, the most commonly used non-lethal storage of biological material at ultra-low temperature is by employing liquid nitrogen. At the temperature of liquid nitrogen (-196°C), almost all the metabolic activities of cells are ceased and the sample can then be preserved in such state for extended periods.

Steps Involved in Cryopreservation

The technique of cryopreservation involves the following steps:

1. Selection of plant material

2. Pre-culture

3. Cryoprotective treatment

4. Freezing and storage

5. Thawing

6. Reculture

Selection of Plant Material

The morphological and physiological conditions of the plant material, prior to freezing, considerably influence its ability to survive freezing at -196°C. Generally, small, richly cytoplasmic and meristematic cells survive better than the larger, highly vacuolated cells. Therefore, cell suspensions should be frequently subcultured and frozen in the late lag phase or exponential phase when the majority of the cells are in the preferred condition. While preservation of cell lines remains useful with respect to *in vitro* production of secondary metabolites, cultured cells are not the ideal system for germplasm storage. Instead, organized structures, such as shoot apices, embryos or young plantlets are preferred. The reasons to shift from cell cultures to organized cultures are as follows:

i.   The genetic instability of cells in long term callus and cell suspension cultures is a very common phenomenon and there is no effective measure to control it so far. Moreover, most of the callus cultures are initiated from non-meristematic cells of the plant body which might exhibit polysomaty. Hence, the cultured cells may exhibit genetic heterogeneity from the very beginning. In contrast, plants raised from shoot apices have generally proved to be true- to-type.

ii. Cultured cells of several important plants do not exhibit totipotency. Moreover, in few cases these cells initially form organs/ embryos and whole plants but this potentiality is often lost after some time in culture. Besides, shoot apices possess a high regeneration ability which is retained in prolonged cultures. Shoot apices are mostly preferred to develop a virus free plants and also for the rapid clonal multiplication.

iii. Haploidy, which is highly unstable in callus and suspension cultures can be maintained through shoot tip culture and axillary-bud proliferation.

iv. The cells of shoot-tip and young embryos are small and meristematic. They ppear to be better suited than larger cells to survive liquid nitrogen (LN) freezing and thawing.

Pre-culture

In several cases, a brief culture of shoot apices for at least 48h at 4°C before freezing has proved beneficial for consistently high frequency of survival of shoot apices after freezing in liquid nitrogen. The other treatments include the application of additives that known to enhance plant stress tolerance, for example ABA, proline, osmoticum (sucrose, mannitol), dimethylsulfoide (DMSO, 1-5%). Sugars acts as osmotically effective agents, although they do not penetrate inside the cells. Dehydration of cells/tissues occurs in the presence of sugars during the preculture, which prevents lethal ice crystal formation during freezing. Proline may act by reducing the level of latent injury to the cells or it may actively participate in recovery metabolism.

Cryoprotective Treatments

There are two potential sources of cell damage during cryopreservation (1) Formation of large ice crystals, inside the cells, leading to rupture of organelle and the cell itself, (2) intracellular concentration of solutes increases to toxic levels before or during freezing as a result of dehydration. Addition of cryoprotectants controls the appearance of ice crystals in cells and protects these cells from the toxic solution effect. Cryoprotectants are categorized as: (a) Penetrating, which exert their protective colligative action, (b) Non-penetrating, which affect through osmotic dehydration. A large number of heterogeneous groups of compounds have been shown to possess cryoprotective properties with different efficiencies, e.g. glycerol, DMSO etc. Cryoprotectant depress both the freezing and super-cooling point of water, i.e. the temperature at which the homogeneous nucleation of ice occurs, thus, retarding the growth of ice crystal formation in cells and protect cells from toxic effect. The cryoprotectants used in cryopreservation are:

a. Alcohols: Ethylene glycol, glycerol, propylene glycol, sorbitol, mannitol

b. Sulphur containing compounds: Amino acids, dimethyl sulphoxide (DMSO), sugar (glucose, saccharose)

c. Polymers: Hydroxyethyl amidon, polyethylene glycol, polyvinyl pyrrolidine

i. Vitrification

At a sufficiently low temperature, highly concentrated aqueous solutions of cryoprotective agents become so viscous that they solidify into an amorphous "glassy" state, without ice crystal formation (crystallization) at practical cooling rates, this phenomenon is called vitrification. The significance of vitrification in cryopreservation of biological materials is that the cells applied with highly concentrated solution of osmotially active compounds, are protected from internal damage from ice crystal formation during freezing. This pretreatment also causes dehydration of cells. The commonly used cryoprotectants are employed for vitrification like DMSO.

ii. Cryoprotective dehydration

If cells are sufficiently dehydrated they may be able to withstand immersion in liquid nitrogen without further application of traditional cryoprotectant mixtures. Dehydration can be achieved by growing the cultures in the presence of high concentration of osmotically active compounds (sugars) and /or air desiccations in laminar-air-flow cabinet or over silica gel. Dehydration reduces the amount of water available for the ice formation.

iii. Encapsulation and dehydration

This involves the encapsulation of tissues in calcium alginate beads which are pre-grown in liquid culture media containing high concentrations of sucrose. The beads are transferred to sterile airflow in a laminar cabinet and desiccated further. After these treatments, the cells are able to withstand exposure to liquid nitrogen without application of chemical cryoprotectants.

Freezing and Storage

The type of crystal water within stored cells is very important for survival of the tissue. Different tissues have different sensitivities for cooling rates. In general, there are three different types of freezing procedures:

i. Rapid freezing

The plant material is placed in vials, liquid nitrogen is poured directly in the vial and dipping the vial into an open flask filled with liquid nitrogen. In this procedure, cooling between -10°C and -70°C occurred at the rate of >1000°C/min. The quicker the freezing is done, smaller the intracellular ice crystals are formed. In combination with desiccation or vitrification pre-treatments, ultra rapid cooling is proved to be the most attractive method for cryopreservation of plant materials. This method has been successfully used for the cryopreservation of shoot-tips, somatic embryos and embryonal axes from zygotic embryos of a number of plant species. The survival rate of cryopreserved tissues by this method is high and when the desiccation pretreatment is applied even the cryoprotectants are not required.

ii. Slow freezing

The tissue is slowly frozen at a slow cooling rate of 0.5-4°C/min from 0 to -100°C, and then transferred to liquid nitrogen. Survival of cells frozen at slow freezing rates may involve some beneficial effects of dehydration, which minimizes the amount of water that freezes intracellularly. Slow cooling permits the flow of water from the cells to the outside, thereby promoting extracellular ice formation instead of intracellular freezing. It is generally agreed that upon extracellular freezing the cytoplasm will be effectively concentrated and plant cells will survive better when adequately dehydrated. This has been successfully employed for cryopreservation of meristems of few plants and has proved especially successful with cells from suspension cultures.

iii. Stepwise freezing

Firstly, the material is cooled gradually (ca 1°C/min) or step-wise (5°C/min) to an optimum intermediate temperature (-30°C to -50°C) for about 30 min, and then rapidly cooled by dipping into liquid nitrogen. The method is highly favorable for freeze preservation of shoot apices and buds. It is equally successful to cells from suspension cultures.

The initial slow freezing reduces the amount of intracellular freezable water by dehydrating the cells. Early in the freezing process ice is formed first outside the cells, and the unfrozen protoplasm of cells loses water due to the vapor pressure deficit between the supercooled protoplasm and the external ice. This initial cooling, thus, acts as another pre-treatment for dehydration of the cells.

Storage

Maintaining the frozen material at the correct temperature is as important as proper freezing itself. Temperatures above -130°C may allow ice-crystal growth inside the cells and, as a result reduce their viability. Long-term storage of the material frozen at -196°C, therefore, requires a liquid nitrogen refrigerator.

Generally, the frozen cells or tissues are immediately kept for storage at temperature ranging from -70°C to -196°C. The storage is ideally done in liquid nitrogen refrigerator at -150°C in the vapor phase or -196°C in the liquid phase. The temperature should be sufficiently low for long term storage of cells to arrest all metabolic activities and to prevent biochemical injury.

Thawing

Rapid thawing of the material frozen at -196°C is achieved by plunging it into water at 37 to 40°C which gives thawing rate of 500-750°C/min. After about 90s, the material is transferred to an ice bath and maintained there until recultured or its viability is tested. The transfer is necessary because the cells might get damage if it is left long in the water bath 37-45°C. Rapid thawing protects the cells from the damaging effects of ice crystal formation, which may occur during slow warming.

Re-culturing

The material after thawing should be washed several times to remove the cryoprotectant which may otherwise be toxic to the cells. A gradual dilution of the cryoprotectant is desirable in-order to avoid any deplasmolysis injury to the cells. The plant material frozen at -196°C may need some special requirements for better survival when re-cultured. Shoot-tips from frozen seedlings of tomato directly developed into plantlets only if the medium was supplemented with GA_3. In its absence, apices callused, followed by the differentiation of adventitious shoots.

References

- Souza, G. R. et al. Three-dimensional Tissue Culture Based on Magnetic Cell Levitation. Nature Nanotechnol.5, 291-296, doi:10.1038/nnano.2010.23 (2010)

- Goldman, Eric, "Wikipedia's Labor Squeeze and its Consequences", Journal on Telecommunications and High Technology Law, 8

- Majchrzak, A.; Wagner, C.; Yates, D. (2006), "Corporate wiki users: results of a survey", Proceedings of the 2006 international symposium on Wikis, Symposium on Wikis, pp. 99–104, doi:10.1145/1149453.1149472, ISBN 1-59593-413-8, retrieved April 25, 2011

- Prestwich, G. D. Simplifying the extracellular matrix for 3-D cell culture and tissue engineering: A pragmatic approach. J. Cell. Biochem.101, 1370-1383, doi:10.1002/jcb.21386 (2007)

- Barsky, Eugene; Giustini, Dean (December 2007), "Introducing Web 2.0: wikis for health librarians" (PDF), Journal of the Canadian Health Libraries Association, 28 (4), pp. 147–150, doi:10.5596/c07-036, ISSN 1708-6892, retrieved November 7, 2011

An Overview of Plant Genetic Engineering

Plant genetic engineering helps in introducing new traits in plants by modifying their DNA. The methods used in this process are gene gun bombardment, microinjection, agrobacterium, electroporation etc. The chapter strategically encompasses and incorporates the major components and key concepts of plant genetic engineering, providing a complete understanding.

Genetic Material

The research at the end of the 19th century had verified Mendelian inheritance and it was also believed that the genetic material is in the chromosome. However, scientists still didn't know the true features of the genetic material. In the early twentieth century, biologists believed that proteins carried genetic information. But the Griffith experiment with *Streptococcus pneumoniae* (1928), Avery, MacLeod and McCarty experiment (1944) on transforming principle and Hershey-Chase experiment (1952) on bacteriophage T2, confirms that DNA is genetic material. Genetic material is the material that determines the inherited characteristics of a functional organism. It has the following properties:

- It must be stable
- It must be capable of being expressed when needed
- It must be capable of accurate replication
- It must be transmitted from parent to progeny without change

The Transforming Principle

In the year 1928, Griffith used two types of strains of pneumonia-causing bacterium, *Streptococcus pneumoniae*, using mice for his experiment. One was S-type (smooth) strain having a polysaccharide coat and produces smooth, shiny colonies on a lab plate. The polysaccharide coat of S-type makes it resistant to the immune system of mice. The other strain, R-type (rough) strain, lacks the coat and produces colonies that look rough and irregular. The R-type lacks the polysaccharide coat and thus it will be destroyed by the immune system of the host. Griffith discovered that there was 'something' that causes the convertion of the R-strain to virulent S- strain.

In the first stage of the Griffith's experiment, he showed that when mouse was injected with S-type strain, mouse died of pneumonia but when injected with R-type strain, mouse lived. The next stage showed that if heat-killed S-type strain was injected to mouse, all mice lived, and this result suggested that the bacteria had been rendered ineffective. The attractive results came with the third part of the experiment, where mice were injected with a mixture of heat-killed S-type strain and non-virulent R-type strains; interestingly all mice developed pneumonia and died. In their blood, Griffith found live bacteria of the deadly S- type. The S strain extract somehow had "transformed" the R- strain bacteria to S- form.

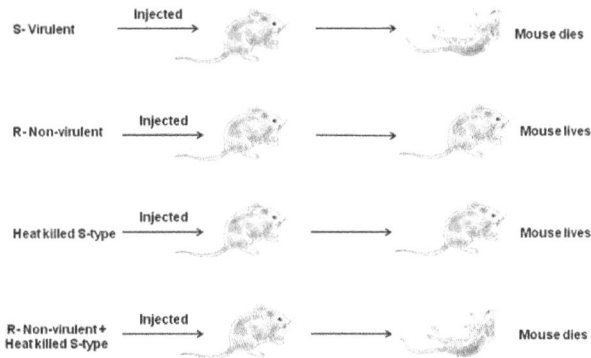

Griffith's experiment discovering the "transforming principle" in *Pneumococcus* bacteria

DNA as the Transforming Principle

Bacteriologists were interested in the difference between the two strains of Streptococci that Frederick Griffith had identified in 1928. The bacteriologists suspected the transforming factor was some kind of protein. The transforming principle could be involved with alcohol, which showed that it was not a carbohydrate, like the polysaccharide coat itself. But Avery and MacCarty observed that proteases, enzymes that degrade proteins, did not destroy the transforming principle. Neither did lipases, enzymes that digest lipids. They found that the transforming substance was rich in nucleic acids, but ribonuclease, which digests RNA, did not inactivate the substance. Avery and McCarty also found that the transforming principle had a high molecular weight. In 1944, Oswald Avery, Colin MacLeod and Maclyn MacCarty showed in their experiments that DNA (not proteins) can transform the properties of cells. They had isolated DNA which was the agent to produce an enduring, heritable change in an organism. Thus, clarifying the chemical nature of the genes and proving that DNA as the "transforming principle" while studying *Streptococcus pneumonia*, bacteria that can cause pneumonia. Until then, biochemists had assumed that deoxyribonucleic acid was a relatively unimportant, structural chemical in chromosomes and that protein, with their greater chemical complexity, transmitted genetic traits.

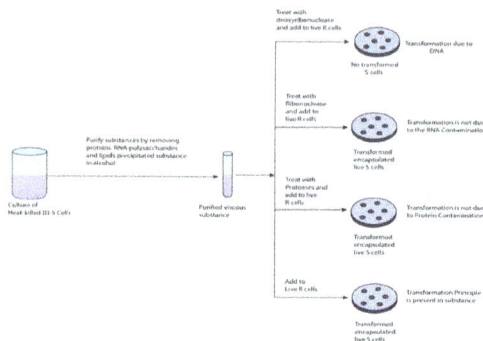

Avery, Macleod and McCarty's experiment demonstrating the
Griffith's transformation

Genetically Modified Crops

Genetically modified crops (GMCs, GM crops, or biotech crops) are plants used in agriculture, the DNA of which has been modified using genetic engineering techniques. In most cases, the aim is to introduce a new trait to the plant which does not occur naturally in the species. Examples in food crops include resistance to certain pests, diseases, or environmental conditions, reduction of spoilage, or resistance to chemical treatments (e.g. resistance to a herbicide), or improving the nutrient profile of the crop. Examples in non-food crops include production of pharmaceutical agents, biofuels, and other industrially useful goods, as well as for bioremediation.

Farmers have widely adopted GM technology. Between 1996 and 2015, the total surface area of land cultivated with GM crops increased by a factor of 100, from 17,000 km² (4.2 million acres) to 1,797,000 km² (444 million acres). 10% of the world's arable land was planted with GM crops in 2010. In the US, by 2014, 94% of the planted area of soybeans, 96% of cotton and 93% of corn were genetically modified varieties. Use of GM crops expanded rapidly in developing countries, with about 18 million farmers growing 54% of worldwide GM crops by 2013. A 2014 meta-analysis concluded that GM technology adoption had reduced chemical pesticide use by 37%, increased crop yields by 22%, and increased farmer profits by 68%. This reduction in pesticide use has been ecologically beneficial, but benefits may be reduced by overuse. Yield gains and pesticide reductions are larger for insect-resistant crops than for herbicide-tolerant crops. Yield and profit gains are higher in developing countries than in developed countries.

There is a scientific consensus that currently available food derived from GM crops poses no greater risk to human health than conventional food, but that each GM food needs to be tested on a case-by-case basis before introduction. Nonetheless, members of the public are much less likely than scientists to perceive GM foods as safe. The legal and regulatory status of GM foods varies by country, with some nations banning or restricting them, and others permitting them with widely differing degrees of regulation.

However, opponents have objected to GM crops on several grounds, including environmental concerns, whether food produced from GM crops is safe, whether GM crops are needed to address the world's food needs, and concerns raised by the fact these organisms are subject to intellectual property law.

Gene Transfer in Nature and Traditional Agriculture

DNA transfers naturally between organisms. Several natural mechanisms allow gene flow across species. These occur in nature on a large scale – for example, it is one mechanism for the development of antibiotic resistance in bacteria. This is facilitated by transposons, retrotransposons, proviruses and other mobile genetic elements that naturally translocate DNA to new loci in a genome. Movement occurs over an evolutionary time scale.

The introduction of foreign germplasm into crops has been achieved by traditional crop breeders by overcoming species barriers. A hybridcereal grain was created in 1875, by crossing wheat and rye. Since then important traits including dwarfing genes and rust resistance have been introduced.Plant tissue culture and deliberate mutations have enabled humans to alter the makeup of plant genomes.

History

The first genetically modified crop plant was produced in 1982, an antibiotic-resistant tobacco plant. The first field trials occurred in France and the USA in 1986, when tobacco plants were engineered for herbicide resistance. In 1987, Plant Genetic Systems (Ghent, Belgium), founded by Marc Van Montagu and Jeff Schell, was the first company to genetically engineer insect-resistant (tobacco) plants by incorporating genes that produced insecticidal proteins from *Bacillus thuringiensis* (Bt).

The People's Republic of China was the first country to allow commercialized transgenic plants, introducing a virus-resistant tobacco in 1992, which was withdrawn in 1997. The first genetically modified crop approved for sale in the U.S., in 1994, was the *FlavrSavr* tomato. It had a longer shelf life, because it took longer to soften after ripening. In 1994, the European Union approved tobacco engineered to be resistant to the herbicide bromoxynil, making it the first commercially genetically engineered crop marketed in Europe.

In 1995, Bt Potato was approved by the US Environmental Protection Agency, making it the country's first pesticide producing crop. In 1995 canola with modified oil composition (Calgene), Bt maize (Ciba-Geigy), bromoxynil-resistant cotton (Calgene), Bt cotton (Monsanto), glyphosate-resistant soybeans (Monsanto), virus-resistant squash (Asgrow), and additional delayed ripening tomatoes (DNAP, Zeneca/Peto, and Monsanto) were approved. As of mid-1996, a total of 35 approvals had been granted to commercially grow 8 transgenic crops and one flower crop (carnation), with 8 different traits in 6 countries plus the EU. In 2000, Vitamin A-enriched golden rice was developed, though as of 2016 it was not yet in commercial production. In 2013 the leaders of the three research teams that first applied genetic engineering to crops, Robert Fraley,

Marc Van Montagu and Mary-Dell Chilton were awarded the World Food Prize for improving the "quality, quantity or availability" of food in the world.

Methods

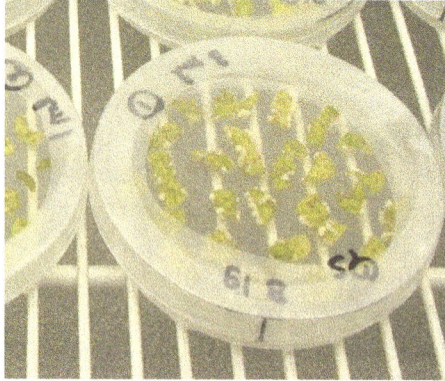

Plants (*Solanum chacoense*) being transformed using agrobacterium

Genetically engineered crops have genes added or removed using genetic engineering techniques, originally including gene guns, electroporation, microinjection and agrobacterium. More recently, CRISPR and TALEN offered much more precise and convenient editing techniques.

Gene guns (also known as biolistics) "shoot" (direct high energy particles or radiations against) target genes into plant cells. It is the most common method. DNA is bound to tiny particles of gold or tungsten which are subsequently shot into plant tissue or single plant cells under high pressure. The accelerated particles penetrate both the cell wall and membranes. The DNA separates from the metal and is integrated into plant DNA inside the nucleus. This method has been applied successfully for many cultivated crops, especially monocots like wheat or maize, for which transformation using *Agrobacterium tumefaciens* has been less successful. The major disadvantage of this procedure is that serious damage can be done to the cellular tissue.

Agrobacterium tumefaciens-mediated transformation is another common technique. Agrobacteria are natural plant parasites, and their natural ability to transfer genes provides another engineering method. To create a suitable environment for themselves, these Agrobacteria insert their genes into plant hosts, resulting in a proliferation of modified plant cells near the soil level (crown gall). The genetic information for tumour growth is encoded on a mobile, circular DNA fragment (plasmid). When *Agrobacterium* infects a plant, it transfers this T-DNA to a random site in the plant genome. When used in genetic engineering the bacterial T-DNA is removed from the bacterial plasmid and replaced with the desired foreign gene. The bacterium is a vector, enabling transportation of foreign genes into plants. This method works especially well for dicotyledonous plants like potatoes, tomatoes, and tobacco. Agrobacteria infection is less successful in crops like wheat and maize.

Electroporation is used when the plant tissue does not contain cell walls. In this technique, "DNA enters the plant cells through miniature pores which are temporarily caused by electric pulses."

Microinjection directly injects the gene into the DNA.

Plant scientists, backed by results of modern comprehensive profiling of crop composition, point out that crops modified using GM techniques are less likely to have unintended changes than are conventionally bred crops.

In research tobacco and *Arabidopsis thaliana* are the most frequently modified plants, due to well-developed transformation methods, easy propagation and well studied genomes. They serve as model organisms for other plant species.

Introducing new genes into plants requires a promoter specific to the area where the gene is to be expressed. For instance, to express a gene only in rice grains and not in leaves, an endosperm-specific promoter is used. The codons of the gene must be optimized for the organism due to codon usage bias.

Types of Modifications

Transgenic maize containing a gene from the bacteria *Bacillus thuringiensis*

Transgenic

Transgenic plants have genes inserted into them that are derived from another species. The inserted genes can come from species within the same kingdom (plant to plant) or between kingdoms (for example, bacteria to plant). In many cases the inserted DNA has to be modified slightly in order to correctly and efficiently express in the host organism. Transgenic plants are used to express proteins like the cry toxins from *B. thuringiensis*, herbicide resistant genes, antibodies and antigens for vaccinations A study led by the European Food Safety Authority (EFSA) found also viral genes in transgenic plants.

Transgenic carrots have been used to produce the drug Taliglucerase alfa which is used to treat Gaucher's disease. In the laboratory, transgenic plants have been modified to increase photosynthesis (currently about 2% at most plants versus the theoretic potential of 9–10%). This is possible by changing the rubisco enzyme (i.e. changing C3 plants into C4 plants), by placing the rubisco in a carboxysome, by adding CO_2 pumps in the cell wall, by changing the leaf form/size. Plants have been engineered to exhibit bioluminescence that may become a sustainable alternative to electric lighting.

Cisgenic

Cisgenic plants are made using genes found within the same species or a closely related one, where conventional plant breeding can occur. Some breeders and scientists argue that cisgenic modification is useful for plants that are difficult to crossbreed by conventional means (such as potatoes), and that plants in the cisgenic category should not require the same regulatory scrutiny as transgenics.

Subgenic

Genetically modified plants can also be developed using gene knockdown or gene knockout to alter the genetic makeup of a plant without incorporating genes from other plants. In 2014, Chinese researcher Gao Caixia filed patents on the creation of a strain of wheat that is resistant to powdery mildew. The strain lacks genes that encode proteins that repress defenses against the mildew. The researchers deleted all three copies of the genes from wheat's hexaploid genome. Gao used the TALENs and CRISPRgene editing tools without adding or changing any other genes. No field trials were immediately planned. The CRISPR technique has also been used to modify white button mushrooms (*Agaricus bisporus*).

Economics

GM food's economic value to farmers is one of its major benefits, including in developing nations. A 2010 study found that Bt corn provided economic benefits of $6.9 billion over the previous 14 years in five Midwestern states. The majority ($4.3 billion) accrued to farmers producing non-Bt corn. This was attributed to European corn borer populations reduced by exposure to Bt corn, leaving fewer to attack conventional corn nearby.Agriculture economists calculated that "world surplus [increased by] $240.3 million for 1996. Of this total, the largest share (59%) went to U.S. farmers. Seed company Monsanto received the next largest share (21%), followed by US consumers (9%), the rest of the world (6%), and the germplasm supplier, Delta & Pine Land Company of Mississippi (5%)."

According to the International Service for the Acquisition of Agri-biotech Applications (ISAAA), in 2014 approximately 18 million farmers grew biotech crops in 28 countries; about 94% of the farmers were resource-poor in developing countries. 53% of the global biotech crop area of 181.5 million hectares was grown in 20 developing countries.

PG Economics comprehensive 2012 study concluded that GM crops increased farm incomes worldwide by $14 billion in 2010, with over half this total going to farmers in developing countries.

Critics challenged the claimed benefits to farmers over the prevalence of biased observers and by the absence of randomized controlled trials. The main Bt crop grown by small farmers in developing countries is cotton. A 2006 review of Bt cotton findings by agricultural economists concluded, "the overall balance sheet, though promising, is mixed. Economic returns are highly variable over years, farm type, and geographical location".

In 2013 the European Academies Science Advisory Council (EASAC) asked the EU to allow the development of agricultural GM technologies to enable more sustainable agriculture, by employing fewer land, water and nutrient resources. EASAC also criticizes the EU's "timeconsuming and expensive regulatory framework" and said that the EU had fallen behind in the adoption of GM technologies.

Participants in agriculture business markets include seed companies, agrochemical companies, distributors, farmers, grain elevators and universities that develop new crops/traits and whose agricultural extensions advise farmers on best practices. According to a 2012 review based on data from the late 1990s and early 2000s, much of the GM crop grown each year is used for livestock feed and increased demand for meat leads to increased demand for GM feedcrops. Feed grain usage as a percentage of total crop production is 70% for corn and more than 90% of oil seed meals such as soybeans. About 65 million metric tons of GM corn grains and about 70 million metric tons of soybean meals derived from GM soybean become feed.

In 2014 the global value of biotech seed was US$15.7 billion; US$11.3 billion (72%) was in industrial countries and US$4.4 billion (28%) was in the developing countries. In 2009, Monsanto had $7.3 billion in sales of seeds and from licensing its technology; DuPont, through its Pioneer subsidiary, was the next biggest company in that market. As of 2009, the overall Roundup line of products including the GM seeds represented about 50% of Monsanto's business.

Some patents on GM traits have expired, allowing the legal development of generic strains that include these traits. For example, generic glyphosate-tolerant GM soybean is now available. Another impact is that traits developed by one vendor can be added to another vendor's proprietary strains, potentially increasing product choice and competition. The patent on the first type of *Roundup Ready* crop that Monsanto produced (soybeans) expired in 2014 and the first harvest of off-patent soybeans occurs in the spring of 2015. Monsanto has broadly licensed the patent to other seed companies that include the glyphosate resistance trait in their seed products. About 150 companies have licensed the technology, including Syngenta and DuPont Pioneer.

Yield

In 2014, the largest review yet concluded that GM crops' effects on farming were positive. The meta-analysis considered all published English-language examinations of the agronomic and economic impacts between 1995 and March 2014 for three major GM crops: soybean, maize, and cotton. The study found that herbicide-tolerant crops have lower production costs, while for insect-resistant crops the reduced pesticide use was offset by higher seed prices, leaving overall production costs about the same.

Yields increased 9% for herbicide tolerance and 25% for insect resistant varieties. Farmers who adopted GM crops made 69% higher profits than those who did not. The review found that GM crops help farmers in developing countries, increasing yields by 14 percentage points.

The researchers considered some studies that were not peer-reviewed, and a few that did not report sample sizes. They attempted to correct for publication bias, by considering sources beyond academic journals. The large data set allowed the study to control for potentially confounding variables such as fertiliser use. Separately, they concluded that the funding source did not influence study results.

Traits

GM crops grown today, or under development, have been modified with various traits. These traits include improved shelf life, disease resistance, stress resistance, herbicide resistance, pest resistance, production of useful goods such as biofuel or drugs, and ability to absorb toxins and for use in bioremediation of pollution.

Recently, research and development has been targeted to enhancement of crops that are locally important in developing countries, such as insect-resistant cowpea for Africa and insect-resistant brinjal (eggplant).

Lifetime

The first genetically modified crop approved for sale in the U.S. was the *FlavrSavr* tomato, which had a longer shelf life. It is no longer on the market.

In November 2014, the USDA approved a GM potato that prevents bruising.

In February 2015 Arctic Apples were approved by the USDA, becoming the first genetically modified apple approved for US sale.Gene silencing was used to reduce the expression of polyphenol oxidase (PPO), thus preventing enzymatic browning of the fruit after it has been sliced open. The trait was added to Granny Smith and Golden Delicious varieties. The trait includes a bacterial antibiotic resistance gene that provides resistance to the antibiotic kanamycin. The genetic engineering involved cultivation in the presence of kanamycin, which allowed only resistant cultivars to survive. Humans consuming apples do not acquire kanamycin resistance, per arcticapple.com. The FDA approved the apples in March 2015.

Nutrition

Edible Oils

Some GM soybeans offer improved oil profiles for processing or healthier eating.Camelina sativa has been modified to produce plants that accumulate high levels of oils similar to fish oils.

Vitamin Enrichment

Golden rice, developed by the International Rice Research Institute (IRRI), provides greater amounts of Vitamin A targeted at reducing Vitamin A deficiency. As of January 2016, golden rice has not yet been grown commercially in any country.

Researchers vitamin-enriched corn derived from South African white corn variety M37W, producing a 169-fold increase in Vitamin A, 6-fold increase in Vitamin C and doubled concentrations of folate. Modified Cavendish bananas express 10-fold the amount of Vitamin A as unmodified varieties.

Toxin Reduction

A genetically modified cassava under development offers lower cyanogenglucosides and enhanced protein and other nutrients (called BioCassava).

In November 2014, the USDA approved a potato, developed by J.R. Simplot Company, that prevents bruising and produces less acrylamide when fried. The modifications prevent natural, harmful proteins from being made via RNA interference. They do not employ genes from non-potato species. The trait was added to the Russet Burbank, Ranger Russet and Atlantic varieties.

Stress Resistance

Plants engineered to tolerate non-biological stressors such as drought,frost, high soil salinity, and nitrogen starvation were in development. In 2011, Monsanto's DroughtGard maize became the first drought-resistant GM crop to receive US marketing approval.

Herbicides

Glyphosate

As of 1999 the most prevalent GM trait was glyphosate-resistance. Glyphosate, (the active ingredient in Roundup and other herbicide products) kills plants by interfering with the shikimate pathway in plants, which is essential for the synthesis of the aromatic amino acids phenylalanine, tyrosine and tryptophan. The shikimate pathway is not present in animals, which instead obtain aromatic amino acids from their diet. More specifically, glyphosate inhibits the enzyme 5-enolpyruvylshikimate-3-phosphate synthase (EPSPS).

This trait was developed because the herbicides used on grain and grass crops at the time were highly toxic and not effective against narrow-leaved weeds. Thus, developing crops that could withstand spraying with glyphosate would both reduce environmental and health risks, and give an agricultural edge to the farmer.

Some micro-organisms have a version of EPSPS that is resistant to glyphosate inhibition. One of these was isolated from an *Agrobacterium* strain CP4 (CP4 EPSPS) that was resistant to glyphosate. The CP4 EPSPS gene was engineered for plant expression by fusing the 5' end of the gene to a chloroplasttransit peptide derived from the petunia EPSPS. This transit peptide was used because it had shown previously an ability to deliver bacterial EPSPS to the chloroplasts of other plants. This CP4 EPSPS gene was cloned and transfected into soybeans.

The plasmid used to move the gene into soybeans was PV-GMGTO4. It contained three bacterial genes, two CP4 EPSPS genes, and a gene encodingbeta-glucuronidase (GUS) from *Escherichia coli* as a marker. The DNA was injected into the soybeans using the particle acceleration method. Soybean cultivar A5403 was used for the transformation.

Bromoxynil

Tobacco plants have been engineered to be resistant to the herbicide bromoxynil.

Glufosinate

Crops have been commercialized that are resistant to the herbicide glufosinate, as well. Crops engineered for resistance to multiple herbicides to allow farmers to use a mixed group of two, three, or four different chemicals are under development to combat growing herbicide resistance.

2,4-D

In October 2014 the US EPA registered Dow's Enlist Duo maize, which is genetically modified to be resistant to both glyphosate and 2,4-D, in six states. Inserting a bacterial aryloxyalkanoate dioxygenase gene, *aad1* makes the corn resistant to 2,4-D. The USDA had approved maize and soybeans with the mutation in September 2014.

Dicamba

Monsanto has requested approval for a stacked strain that is tolerant of both glyphosate and dicamba.

Pest Resistance

Insects

Tobacco, corn, rice and many other crops have been engineered to express genes encoding for insecticidal proteins from Bacillus thuringiensis (Bt). Papaya, potatoes, and squash have been engineered to resist viral pathogens such as cucumber mosaic virus which, despite its name, infects a wide variety of plants. The introduction of Bt crops during the period between 1996 and 2005 has been estimated to have reduced the total volume of insecticide active ingredient use in the United States by over 100 thousand tons. This represents a 19.4% reduction in insecticide use.

In the late 1990s, a genetically modified potato that was resistant to the Colorado potato beetle was withdrawn because major buyers rejected it, fearing consumer opposition.

Viruses

Virus resistant papaya were developed In response to a papaya ringspot virus (PRV) outbreak in Hawaii in the late 1990s. . They incorporate PRV DNA. By 2010, 80% of Hawaiian papaya plants were genetically modified.

Potatoes were engineered for resistance to potato leaf roll virus and Potato virus Y in 1998. Poor sales led to their market withdrawal after three years.

Yellow squash that were resistant to at first two, then three viruses were developed, beginning in the 1990s. The viruses are watermelon, cucumber and zucchini/courgette yellow mosaic. Squash was the second GM crop to be approved by US regulators. The trait was later added to zucchini.

Many strains of corn have been developed in recent years to combat the spread of Maize dwarf mosaic virus, a costly virus that causes stunted growth which is carried in Johnson grass and spread by aphid insect vectors. These strands are commercially available although the resistance is not standard among GM corn variants.

By-products

Drugs

In 2012, the FDA approved the first plant-produced pharmaceutical, a treatment for Gaucher's Disease. Tobacco plants have been modified to produce therapeutic antibodies.

Biofuel

Algae is under development for use in biofuels. Researchers in Singapore were working on GM jatropha for biofuel production.Syngenta has USDA approval to market a maize trademarked Enogen that has been genetically modified to convert its starch to sugar for ethanol. In 2013, the Flemish Institute for Biotechnology was investigating poplar trees genetically engineered to contain less lignin to ease conversion into ethanol. Lignin is the critical limiting factor when using wood to make bio-ethanol because lignin limits the accessibility of cellulosemicrofibrils to depolymerization by enzymes.

Materials

Companies and labs are working on plants that can be used to make bioplastics. Potatoes that produce industrially useful starches have been developed as well. Oilseed can be modified to produce fatty acids for detergents, substitute fuels and petrochemicals.

Bioremediation

Scientists at the University of York developed a weed (*Arabidopsis thaliana*) that contains genes from bacteria that could clean TNT and RDX-explosive soil contaminants in 2011. 16 million hectares in the USA (1.5% of the total surface) are estimated to be contaminated with TNT and RDX. However *A. thaliana* was not tough enough for use on military test grounds. Modifications in 2016 included switchgrass and bentgrass.

Genetically modified plants have been used for bioremediation of contaminated soils. Mercury, selenium and organic pollutants such as polychlorinated biphenyls (PCBs).

Marine environments are especially vulnerable since pollution such as oil spills are not containable. In addition to anthropogenic pollution, millions of tons of petroleum annually enter the marine environment from natural seepages. Despite its toxicity, a considerable fraction of petroleum oil entering marine systems is eliminated by the hydrocarbon-degrading activities of microbial communities. Particularly successful is a recently discovered group of specialists, the so-called hydrocarbonoclastic bacteria (HCCB) that may offer useful genes.

Asexual Reproduction

Crops such as maize reproduce sexually each year. This randomizes which genes get propagated to the next generation, meaning that desirable traits can be lost. To maintain a high-quality crop, some farmers purchase seeds every year. Typically, the seed company maintains two inbred varieties, and crosses them into a hybrid strain that is then sold. Related plants like sorghum and gamma grass are able to perform apomixis, a form of asexual reproduction that keeps the plant's DNA intact. This trait is apparently controlled by a single dominant gene, but traditional breeding has been unsuccessful in creating asexually-reproducing maize. Genetic engineering offers another route to this goal. Successful modification would allow farmers to replant harvested seeds that retain desirable traits, rather than relying on purchased seed.

Crops

Herbicide Tolerance

GMO	Use	Countries approved in	First approved	Notes
Alfalfa	Animal feed	USA	2005	Approval withdrawn in 2007 and then re-approved in 2011
Canola	Cooking oil Margarine Emulsifiers in packaged foods	Australia	2003	
		Canada	1995	
		USA	1995	
Cotton	Fiber Cottonseed oil Animal feed	Argentina	2001	
		Australia	2002	
		Brazil	2008	
		Columbia	2004	
		Costa Rica	2008	
		Mexico	2000	
		Paraguay	2013	
		South Africa	2000	
		USA	1994	
Maize	Animal feed high-fructose corn syrup corn starch	Argentina	1998	
		Brazil	2007	
		Canada	1996	
		Colombia	2007	
		Cuba	2011	
		European Union	1998	Grown in Portugal, Spain, Czech Republic, Slovakia and Romania
		Honduras	2001	
		Paraguay	2012	
		Philippines	2002	
		South Africa	2002	
		USA	1995	
		Uruguay	2003	

GMO	Use	Countries approved in	First approved	Notes
Soy-bean	Animal feed Soybean oil	Argentina	1996	
		Bolivia	2005	
		Brazil	1998	
		Canada	1995	
		Chile	2007	
		Costa Rica	2001	
		Mexico	1996	
		Paraguay	2004	
		South Africa	2001	
		USA	1993	
		Uruguay	1996	
Sugar Beet	Food	Canada	2001	
		USA	1998	Commercialised 2007, production blocked 2010, resumed 2011.

Insect Resistance

GMO	Use	Countries approved in	First approved	Notes
Cotton	Fiber Cottonseed oil Animal feed	Argentina	1998	
		Australia	2003	
		Brazil	2005	
		Burkina Faso	2009	
		China	1997	
		Colombia	2003	
		Costa Rica	2008	
		India	2002	Largest producer of Bt cotton
		Mexico	1996	
		Myanmar	2006	
		Pakistan	2010	
		Paraguay	2007	
		South Africa	1997	
		Sudan	2012	
		USA	1995	

GMO	Use	Countries approved in	First approved	Notes
Eggplant	Food	Bangladesh	2013	12 ha planted on 120 farms in 2014
Maize	Animal feed high-fructose corn syrup corn starch	Argentina	1998	
		Brazil	2005	
		Columbia	2003	
		Mexico	1996	Centre of origin for maize
		Paraguay	2007	
		Philippines	2002	
		South Africa	1997	
		Uruguay	2003	
		USA	1995	
Poplar	Tree	China	1998	543 ha of bt poplar planted in 2014

Other Modified Traits

GMO	Use	Trait	Countries approved in	First approved	Notes
Canola	Cooking oil Margarine Emulsifiers in packaged foods	High laurate canola	Canada	1996	
			USA	1994	
		Phytase production	USA	1998	
Carnation	Ornamental	Delayed senescence	Australia	1995	
			Norway	1998	
		Modified flower colour	Australia	1995	
			Columbia	2000	In 2014 4 ha were grown in greenhouses for export
			European Union	1998	Two events expired 2008, another approved 2007
			Japan	2004	
			Malaysia	2012	For ornamental purposes
			Norway	1997	

GMO	Use	Trait	Countries approved in	First approved	Notes
Maize	Animal feed high-fructose corn syrup corn starch	Increased lysine	Canada	2006	
			USA	2006	
		Drought tolerance	Canada	2010	
			USA	2011	
Papaya	Food	Virus resistance	China	2006	
			USA	1996	Mostly grown in Hawaii
Petunia	Ornamental	Modified flower colour	China	1997	
Potato	Food	Virus resistance	Canada	1999	
			USA	1997	
	Industrial	Modified starch	USA	2014	
Rose	Ornamental	Modified flower colour	Australia	2009	Surrendered renewal
			Colombia	2010	Greenhouse cultivation for export only.
			Japan	2008	
			USA	2011	
Soybean	Animal feed Soybean oil	Increased oleic acidproduction	Argentina	2015	
			Canada	2000	
			USA	1997	
		Stearidonic acid production	Canada	2011	
			USA	2011	
Squash	Food	Virus resistance	USA	1994	
Sugar Cane	Food	Drought tolerance	Indonesia	2013	Environmental certificate only
Tobacco	Cigarettes	Nicotine reduction	USA	2002	

Development

The number of USDA-approved field releases for testing grew from 4 in 1985 to 1,194 in 2002 and averaged around 800 per year thereafter. The number of sites per release and the number of gene constructs (ways that the gene of interest is packaged together with other elements)—have rapidly increased since 2005. Releases with agronomic properties (such as drought resistance) jumped from 1,043 in 2005 to 5,190 in 2013. As of September 2013, about 7,800 releases had been approved for corn, more than 2,200 for soybeans, more than 1,100 for cotton, and about 900 for potatoes. Releases were approved for herbicide tolerance (6,772 releases), insect resistance (4,809), product quality such as flavor or nutrition (4,896), agronomic properties like drought resistance (5,190), and virus/fungal resistance (2,616). The institutions with the most authorized field releases include Monsanto with 6,782, Pioneer/DuPont with 1,405, Syngenta with 565, and USDA's Agricultural Research Service with 370. As of September 2013 USDA had received proposals for releasing GM rice, squash, plum, rose, tobacco, flax and chicory.

Farming Practices

Bt Resistance

Constant exposure to a toxin creates evolutionary pressure for pests resistant to that toxin. Overreliance on glyphosate and a reduction in the diversity of weed management practices allowed the spread of glyphosate resistance in 14 weed species/biotypes in the US.

One method of reducing resistance is the creation of refuges to allow nonresistant organisms to survive and maintain a susceptible population.

To reduce resistance to Bt crops, the 1996 commercialization of transgenic cotton and maize came with a management strategy to prevent insects from becoming resistant. Insect resistance management plans are mandatory for Bt crops. The aim is to encourage a large population of pests so that any (recessive) resistance genes are diluted within the population. Resistance lowers evolutionary fitness in the absence of the stressor (Bt). In refuges, non-resistant strains outcompete resistant ones.

With sufficiently high levels of transgene expression, nearly all of the heterozygotes (S/s), i.e., the largest segment of the pest population carrying a resistance allele, will be killed before maturation, thus preventing transmission of the resistance gene to their progeny. Refuges (i. e., fields of nontransgenic plants) adjacent to transgenic fields increases the likelihood that homozygous resistant (s/s) individuals and any surviving heterozygotes will mate with susceptible (S/S) individuals from the refuge, instead of with other individuals carrying the resistance allele. As a result, the resistance gene frequency in the population remains lower.

Complicating factors can affect the success of the high-dose/refuge strategy. For example, if the temperature is not ideal, thermal stress can lower Bt toxin production and leave the plant more susceptible. More importantly, reduced late-season expression has been documented, possibly resulting from DNA methylation of the promoter. The success of the high-dose/refuge strategy has successfully maintained the value of Bt crops. This success has depended on factors independent of management strategy, including low initial resistance allele frequencies, fitness costs associated with resistance, and the abundance of non-Bt host plants outside the refuges.

Companies that produce Bt seed are introducing strains with multiple Bt proteins. Monsanto did this with Bt cotton in India, where the product was rapidly adopted. Monsanto has also; in an attempt to simplify the process of implementing refuges in fields to comply with Insect Resistance Management(IRM) policies and prevent irresponsible planting practices; begun marketing seed bags with a set proportion of refuge (non-transgenic) seeds mixed in with the Bt seeds being sold. Coined "Refuge-In-a-Bag" (RIB), this practice is intended to increase farmer compliance with refuge requirements and reduce additional labor needed at planting from having separate Bt and refuge seed bags on hand. This strategy is likely to reduce the likelihood of Bt-resistance occurring for corn rootworm, but may increase the risk of resistance for lepidopteran corn pests, such as European corn borer. Increased concerns for resistance with seed mixtures include partially resistant larvae on a Bt plant being able to move to a susceptible plant to survive or cross pollination of refuge pollen on to Bt plants that can lower the amount of Bt expressed in kernels for ear feeding insects.

Herbicide Resistance

Best management practices (BMPs) to control weeds may help delay resistance. BMPs include applying multiple herbicides with different modes of action, rotating crops, planting weed-free seed, scouting fields routinely, cleaning equipment to reduce the transmission of weeds to other fields, and maintaining field borders. The most widely planted GMOs are designed to tolerate herbicides. By 2006 some weed populations had evolved to tolerate some of the same herbicides. Palmer amaranth is a weed that competes with cotton. A native of the southwestern US, it traveled east and was first found resistant to glyphosate in 2006, less than 10 years after GM cotton was introduced.

Plant Protection

Farmers generally use less insecticide when they plant Bt-resistant crops. Insecticide use on corn farms declined from 0.21 pound per planted acre in 1995 to 0.02 pound in 2010. This is consistent with the decline in European corn borer populations as a direct result of Bt corn and cotton. The establishment of minimum refuge requirements helped delay the evolution of Bt resistance. However resistance appears to be developing to some Bt traits in some areas.

Tillage

By leaving at least 30% of crop residue on the soil surface from harvest through planting, conservation tillage reduces soil erosion from wind and water, increases water retention, and reduces soil degradation as well as water and chemical runoff. In addition, conservation tillage reduces the carbon footprint of agriculture. A 2014 review covering 12 states from 1996 to 2006, found that a 1% increase in herbicde-tolerant (HT) soybean adoption leads to a 0.21% increase in conservation tillage and a 0.3% decrease in quality-adjusted herbicide use.

Regulation

The regulation of genetic engineering concerns the approaches taken by governments to assess and manage the risks associated with the development and release of genetically modified crops. There are differences in the regulation of GM crops between countries, with some of the most marked differences occurring between the USA and Europe. Regulation varies in a given country depending on the intended use of each product. For example, a crop not intended for food use is generally not reviewed by authorities responsible for food safety.

Production

In 2013, GM crops were planted in 27 countries; 19 were developing countries and 8 were developed countries. 2013 was the second year in which developing countries grew a majority (54%) of the total GM harvest. 18 million farmers grew GM crops; around 90% were small-holding farmers in developing countries.

Country	2013– GM planted area (million hectares)	Biotech crops
USA	70.1	Maize, Soybean, Cotton, Canola, Sugarbeet, Alfalfa, Papaya, Squash
Brazil	40.3	Soybean, Maize, Cotton
Argentina	24.4	Soybean, Maize, Cotton
India	11.0	Cotton
Canada	10.8	Canola, Maize, Soybean, Sugarbeet
Total	175.2	----

The United States Department of Agriculture (USDA) reports every year on the total area of GMO varieties planted in the United States. According to National Agricultural Statistics Service, the states published in these tables represent 81–86 percent of all corn planted area, 88–90 percent of all soybean planted area, and 81–93 percent of all upland cotton planted area (depending on the year).

Global estimates are produced by the International Service for the Acquisition of Agri-biotech Applications (ISAAA) and can be found in their annual reports, "Global Status of Commercialized Transgenic Crops".

Farmers have widely adopted GM technology. Between 1996 and 2013, the total surface area of land cultivated with GM crops increased by a factor of 100, from 17,000 square kilometers (4,200,000 acres) to 1,750,000 km^2 (432 million acres). 10% of the world's arable land was planted with GM crops in 2010. As of 2011, 11 different transgenic crops were grown commercially on 395 million acres (160 million hectares) in 29 countries such as the USA, Brazil, Argentina, India, Canada, China, Paraguay, Pakistan, South Africa, Uruguay, Bolivia, Australia, Philippines, Myanmar, Burkina Faso, Mexico and Spain. One of the key reasons for this widespread adoption is the perceived economic benefit the technology brings to farmers. For example, the system of planting glyphosate-resistant seed and then applying glyphosate once plants emerged provided farmers with the opportunity to dramatically increase the yield from a given plot of land, since this allowed them to plant rows closer together. Without it, farmers had to plant rows far enough apart to control post-emergent weeds with mechanical tillage. Likewise, using Bt seeds means that farmers do not have to purchase insecticides, and then invest time, fuel, and equipment in applying them. However critics have disputed whether yields are higher and whether chemical use is less, with GM crops.

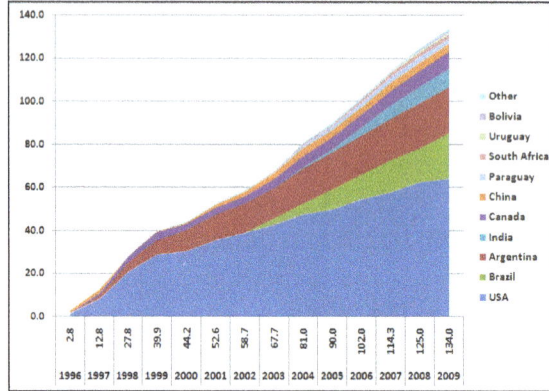

Land area used for genetically modified crops by country (1996–2009), in millions of hectares. In 2011, the land area used was 160 million hectares, or 1.6 million square kilometers.

In the US, by 2014, 94% of the planted area of soybeans, 96% of cotton and 93% of corn were genetically modified varieties. Genetically modified soybeans carried herbicide-tolerant traits only, but maize and cotton carried both herbicide tolerance and insect protection traits (the latter largely Bt protein). These constitute "input-traits" that are aimed to financially benefit the producers, but may have indirect environmental benefits and cost benefits to consumers. The Grocery Manufacturers of America estimated in 2003 that 70–75% of all processed foods in the U.S. contained a GM ingredient.

Europe grows relatively few genetically engineered crops with the exception of Spain, where one fifth of maize is genetically engineered, and smaller amounts in five other countries. The EU had a 'de facto' ban on the approval of new GM crops, from 1999 until 2004. GM crops are now regulated by the EU. In 2015, genetically engineered crops are banned in 38 countries worldwide, 19 of them in Europe. Developing countries grew 54 percent of genetically engineered crops in 2013.

In recent years GM crops expanded rapidly in developing countries. In 2013 approximately 18 million farmers grew 54% of worldwide GM crops in developing countries. 2013's largest increase was in Brazil (403,000 km² versus 368,000 km² in 2012). GM cotton began growing in India in 2002, reaching 110,000 km² in 2013.

According to the 2013 ISAAA brief: "…a total of 36 countries (35 + EU-28) have granted regulatory approvals for biotech crops for food and/or feed use and for environmental release or planting since 1994... a total of 2,833 regulatory approvals involving 27 GM crops and 336 GM events (NB: an "event" is a specific genetic modification in a specific species) have been issued by authorities, of which 1,321 are for food use (direct use or processing), 918 for feed use (direct use or processing) and 599 for environmental release or planting. Japan has the largest number (198), followed by the U.S.A. (165, not including "stacked" events), Canada (146), Mexico (131), South Korea (103), Australia (93), New Zealand (83), European Union (71 including approvals that have expired or under renewal process), Philippines (68), Taiwan (65), Colombia (59), China (55) and South Africa (52). Maize has the largest number (130 events in 27 countries), followed by cotton (49 events in 22 countries), potato (31 events in 10 countries), canola (30 events in 12 countries) and soybean (27 events in 26 countries).

Controversy

GM foods are controversial and the subject of protests, vandalism, referenda, legislation, court action and scientific disputes. The controversies involve consumers, biotechnology companies, governmental regulators, non-governmental organizations and scientists. The key areas are whether GM food should be labeled, the role of government regulators, the effect of GM crops on health and the environment, the effects of pesticide use and resistance, the impact on farmers, and their roles in feeding the world and energy production.

There is a scientific consensus that currently available food derived from GM crops poses no greater risk to human health than conventional food, but that each GM food needs to be tested on a case-by-case basis before introduction. Nonetheless, members of the public are much less likely than scientists to perceive GM foods as safe. The legal and regulatory status of GM foods varies by country, with some nations banning or restricting them, and others permitting them with widely differing degrees of regulation.

No reports of ill effects have been documented in the human population from GM food. Although GMO labeling is required in many countries, the United States Food and Drug Administration does not require labeling, nor does it recognize a distinction between approved GMO and non-GMO foods.

Advocacy groups such as Center for Food Safety, Union of Concerned Scientists, Greenpeace and the World Wildlife Fund claim that risks related to GM food have not been adequately examined and managed, that GMOs are not sufficiently tested and should be labelled, and that regulatory authorities and scientific bodies are too closely tied to industry. Some studies have claimed that genetically modified crops can cause harm; a 2016 review that reanalyzed the data from six of these studies found that their statistical methodologies were flawed and did not demonstrate harm, and said that conclusions about GMO crop safety should be drawn from "the totality of the evidence... instead of far-fetched evidence from single studies".

Hershey–Chase Experiment

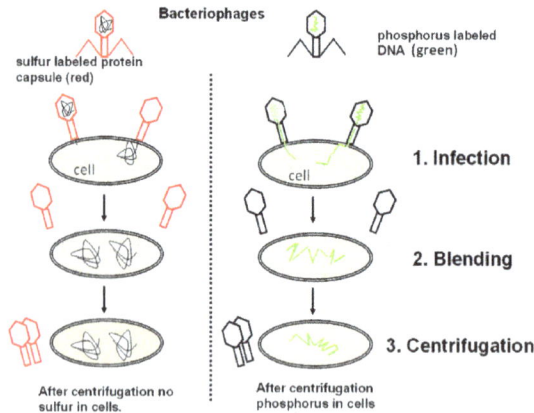

Overview of experiment and observations

The Hershey–Chase experiments were a series of experiments conducted in 1952 by Alfred Hershey and Martha Chase that helped to confirm that DNA is genetic material. While DNA had been known to biologists since 1869, many scientists still assumed at the time that proteins carried the information for inheritance because DNA appeared simpler than proteins. In their experiments, Hershey and Chase showed that when bacteriophages, which are composed of DNA and protein, infect bacteria, their DNA enters the host bacterial cell, but most of their protein does not. Although the results were not conclusive, and Hershey and Chase were cautious in their interpretation, previous, contemporaneous, and subsequent discoveries all served to prove that DNA is the hereditary material.

Hershey shared the 1969 Nobel Prize in Physiology or Medicine with Max Delbrück and Salvador Luria for their "discoveries concerning the genetic structure of viruses."

Historical Background

In the early twentieth century, biologists thought that proteins carried genetic information. This was based on the belief that proteins were more complex than DNA. Phoebus Levene's influential "tetranucleotide hypothesis", which incorrectly proposed that DNA was a repeating set of identical nucleotides, supported this conclusion. The results of the Avery–MacLeod–McCarty experiment, published in 1944, suggested that DNA was the genetic material, but there was still some hesitation within the general scientific community to accept this, which set the stage for the Hershey–Chase experiment.

Hershey and Chase, along with others who had done related experiments, confirmed that DNA was the biomolecule that carried genetic information. Before that, Oswald Avery, Colin MacLeod, and Maclyn McCarty had shown that DNA led to the transformation of one strain of *Streptococcus pneumoniae* to another that was more virulent.

The results of these experiments provided evidence that DNA was the biomolecule that carried genetic information.

Methods and Results

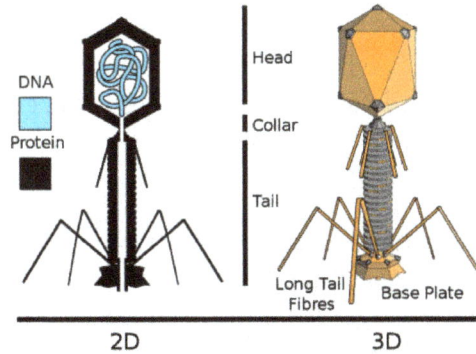

Structural overview of T2 phage

Hershey and Chase needed to be able to examine different parts of the phages they were studying separately, so they needed to isolate the phage subsections. Viruses were known to be composed of a protein shell and DNA, so they chose to uniquely label each with a different elemental isotope. This allowed each to be observed and analyzed separately. Since phosphorus is contained in DNA but not amino acids, radioactive phosphorus-32 was used to label the DNA contained in the T2 phage. Radioactive sulfur-35 was used to label the protein sections of the T2 phage, because sulfur is contained in amino acids but not DNA.

Hershey and Chase inserted the radioactive elements into the bacteriophages by adding the isotopes to separate media within which bacteria were allowed to grow for 4 hours before bacteriophage introduction. When the bacteriophages infected the bacteria, the progeny contained the radioactive isotopes in their structures. This procedure was performed once for the sulfur-labeled phages and once for phosphorus-labeled phages. The labeled progeny were then allowed to infect unlabeled bacteria. The phage coats remained on the outside of the bacteria, while genetic material entered. Disruption of phage from the bacteria by agitation in a blender followed by centrifugation allowed for the separation of the phage coats from the bacteria. These bacteria were lysed to release phage progeny. The progeny of the phages that were originally labeled with ^{32}P remained labeled, while the progeny of the phages originally labeled with ^{35}S were unlabeled. Thus, the Hershey–Chase experiment helped confirm that DNA, not protein, is the genetic material.

Hershey and Chase showed that the introduction of deoxyribonuclease (referred to as DNase), an enzyme that breaks down DNA, into a solution containing the labeled bacteriophages did not introduce any ^{32}P into the solution. This demonstrated that the phage is resistant to the enzyme while intact. Additionally, they were able to plasmolyze the bacteriophages so that they went into osmotic shock, which effectively created a solution containing most of the ^{32}P and a heavier solution containing structures called "ghosts" that contained the ^{35}S and the protein coat of the virus. It was found that these "ghosts" could adsorb to bacteria that were susceptible to T2, although they contained no DNA and were simply the remains of the original bacterial capsule. They concluded that the protein protected the DNA from DNAse, but that once the two were separated and the phage was inactivated, the DNAse could hydrolyze the phage DNA. However, it subsequently became clear that in some viruses, RNA is the genetic material.

Experiment and Conclusions

Hershey and Chase were also able to prove that the DNA from the phage is inserted into the bacteria shortly after the virus attaches to its host. Using a high speed blender they were able to force the bacteriophages from the bacterial cells after adsorption. The lack of ^{32}P labeled DNA remaining in the solution after the bacteriophages had been allowed to adsorb to the bacteria showed that the phage DNA was transferred into the bacterial cell. The presence of almost all the radioactive ^{35}S in the solution showed that the protein coat that protects the DNA before adsorption stayed outside the cell.

Hershey and Chase concluded that DNA, not protein, was the genetic material. They determined that a protective protein coat was formed around the bacteriophage, but that the internal DNA is what conferred its ability to produce progeny inside a bacterium. They showed that, in growth, protein has no function, while DNA has some function. They determined this from the amount of radioactive material remaining outside of the cell. Only 20% of the ^{32}P remained outside the cell, demonstrating that it was incorporated with DNA in the cell's genetic material. All of the ^{35}S in the protein coats remained outside the cell, showing it was not incorporated into the cell, and that protein is not the genetic material.

Hershey and Chase's experiment concluded that little sulfur containing material entered the bacterial cell. However no specific conclusions can be made regarding whether material that is sulfur-free enters the bacterial cell after phage adsorption. Further research was necessary to conclude that it was solely bacteriophages' DNA that entered the cell and not a combination of protein and DNA where the protein did not contain any sulfur.

Discussion

Confirmation

Hershey and Chase concluded that protein was not likely to be the hereditary genetic material. However, they did not make any conclusions regarding the specific function of DNA as hereditary material, and only said that it must have some undefined role.

Confirmation and clarity came a year later in 1953, when James D. Watson and Francis Crick correctly hypothesized, in their journal article "Molecular Structure of Nucleic Acids: A Structure for Deoxyribose Nucleic Acid", the double helix structure of DNA, and suggested the copying mechanism by which DNA functions as hereditary material. Furthermore, Watson and Crick suggested that DNA, the genetic material, is responsible for the synthesis of the thousands of proteins found in cells. They had made this proposal based on the structural similarity that exists between the two macromolecules, that is, both protein and DNA are linear sequences of amino acids and nucleotides respectively.

Other Experiments

Once the Hershey–Chase experiment was published, the scientific community generally acknowledged that DNA was the genetic code material. This discovery led to a more detailed investigation of DNA to determine its composition as well as its 3D structure. Using X-ray crystallography, the structure of DNA was discovered by James Watson and Francis Crick with the help of previously documented experimental evidence by Maurice Wilkins and Rosalind Franklin. Knowledge of the structure of DNA led scientists to examine the nature of genetic coding and, in turn, understand the process of protein synthesis. George Gamow proposed that the genetic code was composed of sequences of three DNA base pairs known as triplets or codons which represent one of the twenty amino acids. Genetic coding helped researchers to understand the mechanism of gene expression, the process by which information from a gene is used in protein synthesis. Since then, much research has been conducted to modulate steps in the gene expression process. These steps include transcription, RNA splicing, translation, and post-translational modification which are used to control the chemical and structural nature of proteins. Moreover, genetic engineering gives engineers the ability to directly manipulate the genetic materials of organisms using recombinant DNA techniques. The first recombinant DNA molecule was created by Paul Berg in 1972 when he combined DNA from the monkey virus SV40 with that of the lambda virus.

Experiments on hereditary material during the time of the Hershey-Chase Experiment often used bacteriophages as a model organism. Bacteriophages lend themselves to experiments on hereditary material because they incorporate their genetic material into their host cell's genetic material (making them useful tools), they multiply quickly, and they are easily collected by researchers.

Legacy

The Hershey–Chase experiment, its predecessors, such as the Avery–MacLeod–McCarty experiment, and successors served to unequivocally establish that hereditary information was carried by DNA. This finding has numerous applications in forensic science, crime investigation and genealogy. It provided the background knowledge for further applications in DNA forensics, where DNA fingerprinting uses data originating from DNA, not protein sources, to deduce genetic variation.

Hershey and Chase in their experiments used the T2 bacteriophage as the vehicle for delivering genetic material. Like all bacterial viruses, T2 is comprised of only a protein-based outer wall and a DNA core, its simple structure making it the perfect research candidate. The virus protein contains sulfur but no phosphorus and the virus DNA contains phosphorus but no sulfur. They tagged the T2 phage DNA with radioactive phosphorous (P^{32}) and proteins with radioactive sulfur (S^{35}). The researchers could track the location of DNA and protein according to the radiation concentrations. Then they allowed the tagged phages for infection to *E.coli*. After introducing phage culture to the bacterial sample, it was agitated in blender to brutally disturb the infected bacteria, causing the protein shells to detach from their hosts. Then, for separation of bacterium from the phages and proteins, they used a centrifuge. Once the separation was complete, they measured the radiation concentrations in the *E.coli* cells and the protein shells. The most of the P^{32} label appeared in large quantities within the bacterial sample, demonstrating that DNA was transferred from the bacteriophage to the host organism whereas most of the S^{35} label had remained outside of the cells. Further, despite the protein shells were detached, reproduction of the phage was taking place and the virus was still copied in each of the host cells. This, suggested that the proteins shell itself was not necessary to the replication process following the initial insertion of genetic material figure.

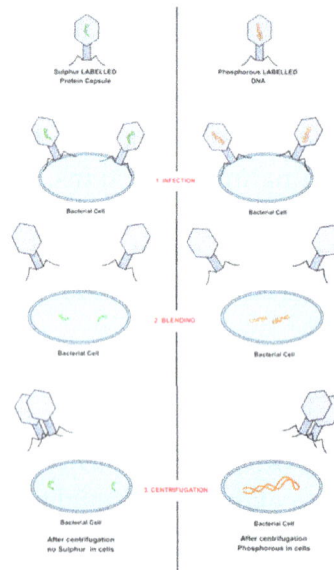

Hersey and Chase's experiment with T2 Bacteriophase in *E. coili*

So from previous explanation we can define the genetic material "*The genetic material of a cell or a plant refers to those materials found in the nucleus, mitochondria and chloroplast, which play a fundamental role in determining the structure and nature of cell substances, and capable of self-propagating and variation. It can be a gene, a part of a gene, a group of genes, a DNA molecule, a fragment of DNA, a group of DNA molecules, or the entire genome of an organism.*"

DNA

Deoxyribonucleic Acid, known as DNA, is the genetic material found in the cells of nearly all living organisms. DNA is the fundamental building blocks of life. Nearly every cell (with a nucleus) in a person's body has the same DNA. Most DNA is located in the cell nucleus (nuclear DNA), but DNA can also be found in the mitochondria (mitochondrial DNA or mt-DNA) and in chloroplast (chloroplast DNA or ctDNA). In 1929 Phoebus Levene at the Rockefeller Institute identified the components that make up a DNA Molecule. The information in DNA is made up of four bases which combine to form chains. These bases include two purines (Adenine and Guanine) and two pyrimidines (Cytosine and Thymine). These are commonly referred to as A, G, C and T, respectively. Each base is attached to a Sugar (S) molecule and a Phosphate (P) molecule. Sugar and phosphate are back bone of nucleotides.Together, a base and a sugar are called a nucleoside. Together, a base, sugar, and phosphate are called a nucleotide. Nucleotides are arranged in two long strands that form a spiral called a double helix. The structure of the double helix is somewhat like a ladder, with the base pairs forming the ladder's rungs and the sugar and phosphate molecules forming the vertical side pieces of the ladder. He showed that the components of DNA were linked in the order phosphate-sugar-base. He called each of these units a nucleotide and suggested the DNA molecule consisted of a string of nucleotide units linked together through the phosphate groups, which are the 'backbone' of the molecule. However Levene thought the chain was short and that the bases repeated in the same fixed order. Torbjorn Caspersson and Einar Hammersten showed that DNA was a polymer. This was only accepted after the structure of DNA was elucidated by James D. Watson and Francis Crick in their 1953 Nature publication. Watson and Crick proposed the central dogma of molecular biology in 1957, describing the process whereby proteins are produced from nucleic DNA. In 1962 Watson, Crick, and Maurice Wilkins jointly received the Nobel Prize for their determination of the structure of DNA. The number of purine bases in DNA is equal to the number of pyrimidines. This is due to the law of complimentary base pairing; where Thymine (T) can only pair with Adenine (A), and Guanine (G) can only pair with Cytosine (C). Knowing this rule, we could predict the base sequence of one DNA strand if we knew the sequence of bases in the complimentary strand.

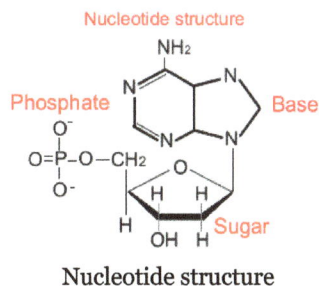

Nucleotide structure

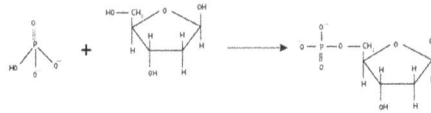

Sugar phosphate backbone of common nucleotides

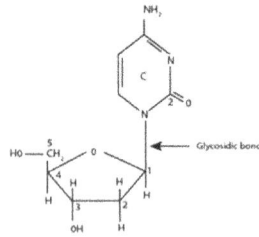

Nucleosides (C=Cytosine)

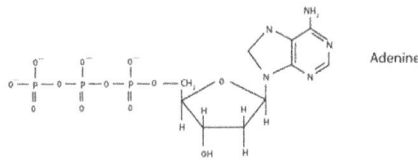

2'- Deoxyribonucleotide

Nucleotides

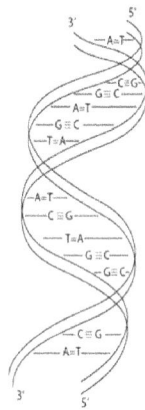

DNA double helix structure

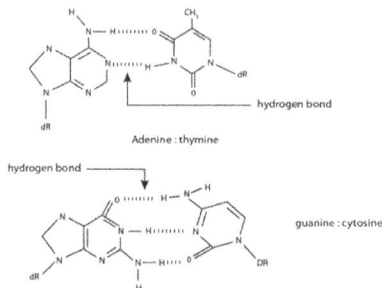

DNA Base pairing by hydrogen bond

The endosymbiotic theory concerns the origins of mitochondria and plastids (e.g. chloroplasts), which are organelles of eukaryotic cells. According to this theory, these organelles originated as separate prokaryotic organisms that were taken inside the cell as endosymbionts. Mitochondria developed from proteobacteria (in particular, Rickettsiales or close relatives) and chloroplasts from cyanobacteria. Mitochondrial and chloroplast genomes do not contain a full set of housekeeping genes, and lack many that other descendants of their speculative ancestors share, there must have been a loss of genes. However, some of these genes likely migrated to the nucleus, where analogues of these genes are now found.

Chloroplast Genome

The chloroplast is the green plastid in land plants, algae and some protists. As the site in the cell where photosynthesis takes place, chloroplasts are responsible for much of the world's primary productivity, making chloroplasts essential to the lives of plants and animals alike. Agriculture, animal farming, and fossil fuels such as coal and oil are all "products" of photosynthesis that took place in chloroplasts. Other important activities that occur in chloroplasts (and several non-photosynthetic plastid types) include the production of starch, certain amino acids and lipids, some of the colorful pigments in flowers, and some key aspects of sulfur and nitrogen metabolism. The interactions between plastid and nuclear encoded transcription and translation process is elaborated in figure. All plastids considered to date contain their own DNA, which is actually a reduced "genome" derived from a cyanobacterial ancestor that was captured early in the evolution of the eukaryotic cell. The chloroplast genome encodes for all the rRNA & tRNA species required for protein synthesis. The ribosomes contain two small rRNAs in addition to the major species. The chloroplast genome codes for ~50 proteins, including RNA polymerase & some ribosomal proteins. Again the rule is that organelle genes are transcribed & translated the apparatus of the organelle. The chloroplast genome of the higher plants varies in length, but displays a characteristic landmark. It has a lengthening sequence, 10-24kb depending on the plants, that is present in two identical copies as an inverted repeat (Gene that are coded within the inverted repeats are present in two copies per genome & include the rRNA genes).

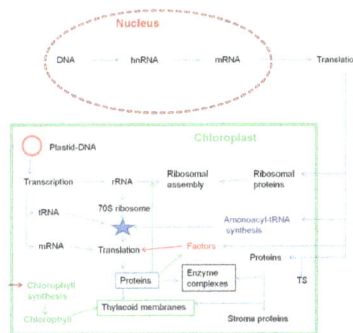

Model of the interactions between plastid and nuclear encoded transcription and translation products. TS: transit sequence: a N-terminal section of the polypeptide chain, essential for the penetration of the polypeptide across the membrane, subsequently being cleaved off proteolytically.

Mitochondrial DNA

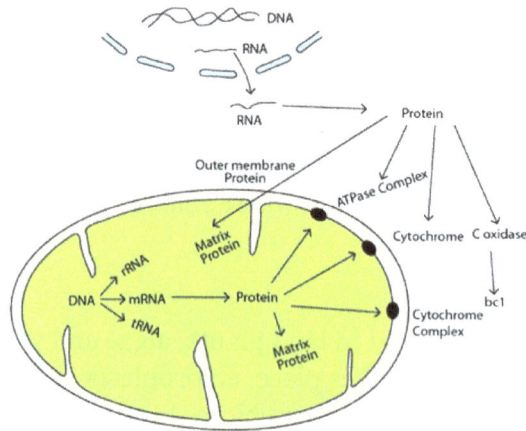

Mitochondrial genome functions

Mitochondrial DNA (mtDNA) is DNA that is present in Mitochondria. Mitochondrion is the part of organic cells that produce most of the cellular energy by converting organic materials mode into Adenosine Tri-phosphate (ATP) mode via the process of oxidative phosphorylation. The details of mitochondrial functions are elaborated in figure. Typically nuclear DNA determines the function of a cell; however mitochondria have their own DNA and are assumed to have evolved separately (Endosymbiotic theory). Mitochondria have their own genome, usually multiple copies in one mitochondrion, in circular form, located in several nucleoid regions, with no histone association (naked). Mitochondrial genome size varies with organism to organism, plants have mitochondrial average 150-200 kb, but human mitochondria genome is only 16 kb. Mitochondrial DNA encodes enzymes required for oxidative phosphorylation and mitochondrial electron transfer. A cell can have different types of mitochondria (heteroplasmy) or same type of mitochondria (homoplasmy). Mitochondrial DNA analysis is helpful in forensic cases in which nuclear DNA is insufficient for short tandem repeat (STR) typing. Shed body, head, and pubic hairs with no cellular material (hair follicle) attached to the root bulb and aged skeletal remains are the samples most commonly analyzed for mtDNA because nuclear DNA is not recoverable from these tissues. Usually a cell has hundreds or thousands of mitochondria which can occupy up to 25% of the cell's cytoplasm, and each mitochondrion contains 1-10 mtDNA molecules. The high copy number of mtDNA molecules found in each cell is one reason why mtDNA is recoverable from hairs and old skeletal remains.

Restriction Enzyme

A restriction enzyme or restriction endonuclease is an enzyme that cuts DNA at or near specific recognition nucleotide sequences known as restriction sites. Restriction enzymes are commonly classified into four types, which differ in their structure and whether they cut their DNA substrate at their recognition site, or if the recognition and cleavage sites are separate from one another. To cut DNA, all restriction enzymes make two incisions, once through each sugar-phosphate backbone (i.e. each strand) of the DNA double helix.

These enzymes are found in bacteria and archaea and provide a defense mechanism against invading viruses. Inside a prokaryote, the restriction enzymes selectively cut up *foreign* DNA in a process called *restriction*; meanwhile, host DNA is protected by a modification enzyme (a methyltransferase) that modifies the prokaryotic DNA and blocks cleavage. Together, these two processes form the restriction modification system.

Over 3000 restriction enzymes have been studied in detail, and more than 600 of these are available commercially. These enzymes are routinely used for DNA modification in laboratories, and are a vital tool in molecular cloning.

History

The term restriction enzyme originated from the studies of phage λ and the phenomenon of host-controlled restriction and modification of a bacterial virus. The phenomenon was first identified in work done in the laboratories of Salvador Luria and Giuseppe Bertani in early 1950s. It was found that, for a bacteriophage λ that can grow well in one strain of *Escherichia coli*, for example *E. coli* C, when grown in another strain, for example *E. coli* K, its yields can drop significantly, by as much as 3-5 orders of magnitude. The host cell, in this example *E. coli* K, is known as the restricting host and appears to have the ability to reduce the biological activity of the phage λ. If a phage becomes established in one strain, the ability of that phage to grow also becomes restricted in other strains. In the 1960s, it was shown in work done in the laboratories of Werner Arber and Matthew Meselson that the restriction is caused by an enzymatic cleavage of the phage DNA, and the enzyme involved was therefore termed a restriction enzyme.

The restriction enzymes studied by Arber and Meselson were type I restriction enzymes, which cleave DNA randomly away from the recognition site. In 1970, Hamilton O. Smith, Thomas Kelly and Kent Wilcox isolated and characterized the first type II restriction enzyme, *Hind*II, from the bacterium *Haemophilus influenzae*. Restriction enzymes of this type are more useful for laboratory work as they cleave DNA at the site of their recognition sequence. Later, Daniel Nathans and Kathleen Danna showed that cleavage of simian virus 40 (SV40) DNA by restriction enzymes yields specific fragments that can be separated using polyacrylamide gel electrophoresis, thus showing that restriction enzymes can also be used for mapping DNA. For their work in the discovery and characterization of restriction enzymes, the 1978 Nobel Prize for Physiology or Medicine was awarded to Werner Arber, Daniel Nathans, and Hamilton O. Smith. The discovery of restriction enzymes allows DNA to be manipulated, leading to the development of recombinant DNA technology that has many applications, for example, allowing the large scale production of proteins such as human insulin used by diabetics.

Restriction enzymes likely evolved from a common ancestor and became widespread via horizontal gene transfer. In addition, there is mounting evidence that restriction endonucleases evolved as a selfish genetic element.

Recognition Site

$$5'\ldots GAT\ |\ ATC\ldots 3'$$
$$3'\ldots CTA\ |\ TAG\ldots 5'$$

A palindromic recognition site reads the same on the reverse strand as it does on the forward strand when both are read in the same orientation

Restriction enzymes recognize a specific sequence of nucleotides and produce a double-stranded cut in the DNA. The recognition sequences can also be classified by the number of bases in its recognition site, usually between 4 and 8 bases, and the amount of bases in the sequence will determine how often the site will appear by chance in any given genome, e.g., a 4-base pair sequence would theoretically occur once every 4^4 or 256bp, 6 bases, 4^6 or 4,096bp, and 8 bases would be 4^8 or 65,536bp. Many of them are palindromic, meaning the base sequence reads the same backwards and forwards. In theory, there are two types of palindromic sequences that can be possible in DNA. The *mirror-like* palindrome is similar to those found in ordinary text, in which a sequence reads the same forward and backward on a single strand of DNA, as in GTAATG. The *inverted repeat* palindrome is also a sequence that reads the same forward and backward, but the forward and backward sequences are found in complementary DNA strands (i.e., of double-stranded DNA), as in GTATAC (GTATAC being complementary to CATATG). Inverted repeat palindromes are more common and have greater biological importance than mirror-like palindromes.

*Eco*RI digestion produces "sticky" ends,

G|A A T T C
C T T A A|G

whereas SmaI restriction enzyme cleavage produces "blunt" ends:

C C C|G G G
G G G|C C C

Recognition sequences in DNA differ for each restriction enzyme, producing differences in the length, sequence and strand orientation (5' end or 3' end) of a sticky-end "overhang" of an enzyme restriction.

Different restriction enzymes that recognize the same sequence are known as neoschizomers. These often cleave in different locales of the sequence. Different enzymes that recognize and cleave in the same location are known as isoschizomers.

Types

Naturally occurring restriction endonucleases are categorized into four groups (Types I, II III, and IV) based on their composition and enzyme cofactor requirements, the nature of their target sequence, and the position of their DNA cleavage site relative to the target sequence. DNA sequence analyses of restriction enzymes however show great variations, indicating that there are more than four types. All types of enzymes recognize specific short DNA sequences and carry out the endonucleolytic cleavage of DNA to give specific fragments with terminal 5'-phosphates. They differ in their recognition sequence, subunit composition, cleavage position, and cofactor requirements, as summarised below:

- Type I enzymes (EC3.1.21.3) cleave at sites remote from a recognition site; require both ATP and S-adenosyl-L-methionine to function; multifunctional protein with both restriction and methylase (EC2.1.1.72) activities.

- Type II enzymes (EC3.1.21.4) cleave within or at short specific distances from a recognition site; most require magnesium; single function (restriction) enzymes independent of methylase.

- Type III enzymes (EC3.1.21.5) cleave at sites a short distance from a recognition site; require ATP (but do not hydrolyse it); S-adenosyl-L-methionine stimulates the reaction but is not required; exist as part of a complex with a modification methylase (EC2.1.1.72).

- Type IV enzymes target modified DNA, e.g. methylated, hydroxymethylated and glucosyl-hydroxymethylated DNA

Type l

Type I restriction enzymes were the first to be identified and were first identified in two different strains (K-12 and B) of *E. coli*. These enzymes cut at a site that differs, and is a random distance (at least 1000 bp) away, from their recognition site. Cleavage at these random sites follows a process of DNA translocation, which shows that these enzymes are also molecular motors. The recognition site is asymmetrical and is composed of two specific portions—one containing 3–4 nucleotides, and another containing 4–5 nucleotides—separated by a non-specific spacer of about 6–8 nucleotides. These enzymes are multifunctional and are capable of both restriction and modification activities, depending upon the methylation status of the target DNA. The cofactors S-Adenosyl methionine (AdoMet), hydrolyzed adenosine triphosphate (ATP), and magnesium (Mg^{2+}) ions, are required for their full activity. Type I restriction enzymes possess three subunits called HsdR, HsdM, and HsdS; HsdR is required for restriction; HsdM is necessary for adding methyl groups to host DNA (methyltransferase activity), and HsdS is important for specificity of the recognition (DNA-binding) site in addition to both restriction (DNA cleavage) and modification (DNA methyltransferase) activity.

Type II

Typical type II restriction enzymes differ from type I restriction enzymes in several ways. They form homodimers, with recognition sites that are usually undivided and palindromic and 4–8 nucleotides in length. They recognize and cleave DNA at the same site, and they do not use ATP or AdoMet for their activity—they usually require only Mg^{2+} as a cofactor. These are the most commonly available and used restriction enzymes. In the 1990s and early 2000s, new enzymes from this family were discovered that did not follow all the classical criteria of this enzyme class, and new subfamily nomenclature was developed to divide this large family into subcategories based on deviations from typical characteristics of type II enzymes. These subgroups are defined using a letter suffix.

Type IIB restriction enzymes (e.g., BcgI and BplI) are multimers, containing more than one subunit. They cleave DNA on both sides of their recognition to cut out the recognition site. They require both AdoMet and Mg^{2+} cofactors. Type IIE restriction endonucleases (e.g., NaeI) cleave DNA following interaction with two copies of their recognition sequence. One recognition site acts as the target for cleavage, while the other acts as an allosteric effector that speeds up or improves the efficiency of enzyme cleavage. Similar to type IIE enzymes, type IIF restriction endonucleases (e.g. NgoMIV) interact with two copies of their recognition sequence but cleave both sequences at the same time. Type IIG restriction endonucleases (e.g., Eco57I) do have a single subunit, like classical Type II restriction enzymes, but require the cofactor AdoMet to be active. Type IIM restriction endonucleases, such as DpnI, are able to recognize and cut methylated DNA. Type IIS restriction endonucleases (e.g., *Fok*I) cleave DNA at a defined distance from their non-palindromic asymmetric recognition sites; this characteristic is widely used to perform in-vitro cloning techniques such as Golden Gate cloning. These enzymes may function as dimers. Similarly, Type IIT restriction enzymes (e.g., Bpu10I and BslI) are composed of two different subunits. Some recognize palindromic sequences while others have asymmetric recognition sites.

Type III

Type III restriction enzymes (e.g., EcoP15) recognize two separate non-palindromic sequences that are inversely oriented. They cut DNA about 20–30 base pairs after the recognition site. These enzymes contain more than one subunit and require AdoMet and ATP cofactors for their roles in DNA methylation and restriction, respectively. They are components of prokaryotic DNA restriction-modification mechanisms that protect the organism against invading foreign DNA. Type III enzymes are hetero-oligomeric, multifunctional proteins composed of two subunits, Res and Mod. The Mod subunit recognises the DNA sequence specific for the system and is a modification methyltransferase; as such, it is functionally equivalent to the M and S subunits of type I restriction endonuclease. Res is required for restriction, although it has no enzymatic activity on its own. Type III enzymes recognise short 5–6 bp-long asymmetric DNA sequences and cleave 25–27 bp downstream to leave short, single-stranded 5' protrusions. They require the presence of two inversely oriented unmethylated recognition sites for restriction to occur. These enzymes methylate only one strand of the DNA, at the N-6 position of adenosyl residues, so newly replicated DNA will have only one strand methylated, which is sufficient to protect against restriction. Type III enzymes belong to the beta-subfamily of N6 adenine methyltransferases, containing the nine motifs that characterise this family, including motif I, the AdoMet binding pocket (FXGXG), and motif IV, the catalytic region (S/D/N (PP) Y/F).

Type IV

Type IV enzymes recognize modified, typically methylated DNA and are exemplified by the McrBC and Mrr systems of *E. coli*.

Type V

Type V restriction enzymes (e.g., the cas9-gRNA complex from CRISPRs) utilize guide RNAs to target specific non-palindromic sequences found on invading organisms. They can cut DNA of variable length, provided that a suitable guide RNA is provided. The flexibility and ease of use of these enzymes make them promising for future genetic engineering applications.

Artificial Restriction Enzymes

Artificial restriction enzymes can be generated by fusing a natural or engineered DNA binding domain to a nuclease domain (often the cleavage domain of the type IIS restriction enzyme *Fok*I). Such artificial restriction enzymes can target large DNA sites (up to 36 bp) and can be engineered to bind to desired DNA sequences.Zinc finger nucleases are the most commonly used artificial restriction enzymes and are generally used in genetic engineering applications, but can also be used for more standard gene cloning applications. Other artificial restriction enzymes are based on the DNA binding domain of TAL effectors.

In 2013 CRISPR-Cas9 are announced. For more detail, read CRISPR(Clustered regularly interspaced short palindromic repeats).

In 2017 a group in Illinois announced using an Argonaute protein taken from Pyrococcus furiosus (PfAgo) along with guide DNA to edit DNA as artificial restriction enzymes.

Nomenclature

Derivation of the *Eco*RI name		
Abbreviation	Meaning	Description
E	*Escherichia*	genus
co	*coli*	specific epithet
R	RY13	strain
I	First identi-fied	order of identification in the bacterium

Since their discovery in the 1970s, many restriction enzymes have been identified; for example, more than 3500 different Type II restriction enzymes have been characterized. Each enzyme is named after the bacterium from which it was isolated, using a naming system based on bacterial genus, species and strain. For example, the name of the *Eco*RI restriction enzyme was derived as shown in the box.

Applications

Isolated restriction enzymes are used to manipulate DNA for different scientific applications.

They are used to assist insertion of genes into plasmidvectors during gene cloning and protein production experiments. For optimal use, plasmids that are commonly used for gene cloning are modified to include a short *polylinker* sequence (called the multiple cloning site, or MCS) rich in restriction enzyme recognition sequences. This allows flexibility when inserting gene fragments into the plasmid vector; restriction sites contained naturally within genes influence the choice of endonuclease for digesting the DNA, since it is necessary to avoid restriction of wanted DNA while intentionally cutting the ends of the DNA. To clone a gene fragment into a vector, both plasmid DNA and gene insert are typically cut with the same restriction enzymes, and then glued together with the assistance of an enzyme known as a DNA ligase.

Restriction enzymes can also be used to distinguish gene alleles by specifically recognizing single base changes in DNA known as single nucleotide polymorphisms (SNPs). This is however only possible if a SNP alters the restriction site present in the allele. In this method, the restriction enzyme can be used to genotype a DNA sample without the need for expensive gene sequencing. The sample is first digested with the restriction enzyme to generate DNA fragments, and then the different sized fragments separated by gel electrophoresis. In general, alleles with correct restriction sites will generate two visible bands of DNA on the gel, and those with altered restriction sites will not be cut and will generate only a single band. A DNA map by restriction digest can also be generated that can give the relative positions of the genes. The different lengths of DNA generated by restriction digest also produce a specific pattern of bands after gel electrophoresis, and can be used for DNA fingerprinting.

In a similar manner, restriction enzymes are used to digest genomic DNA for gene analysis by Southern blot. This technique allows researchers to identify how many copies (or paralogues) of a gene are present in the genome of one individual, or how many gene mutations (polymorphisms) have occurred within a population. The latter example is called restriction fragment length polymorphism (RFLP).

Artificial restriction enzymes created by linking the *Fok*I DNA cleavage domain with an array of DNA binding proteins or zinc finger arrays, denoted zinc finger nucleases (ZFN), are a powerful tool for host genome editing due to their enhanced sequence specificity. ZFN work in pairs, their dimerization being mediated in-situ through the *Fok*I domain. Each zinc finger array (ZFA) is capable of recognizing 9–12 base pairs, making for 18–24 for the pair. A 5–7 bp spacer between the cleavage sites further enhances the specificity of ZFN, making them a safe and more precise tool that can be applied in humans. A recent Phase I clinical trial of ZFN for the targeted abolition of the CCR5 co-receptor for HIV-1 has been undertaken.

Others have proposed using the bacteria R-M system as a model for devising human anti-viral gene or genomic vaccines and therapies since the RM system serves an innate defense-role in bacteria by restricting tropism by bacteriophages. There is research on REases and ZFN that can cleave the DNA of various human viruses, including HSV-2, high-risk HPVs and HIV-1, with the ultimate goal of inducing target mutagenesis and aberrations of human-infecting viruses. Interestingly, the human genome already contains remnants of retroviral genomes that have been inactivated and harnessed for self-gain. Indeed, the mechanisms for silencing active L1 genomic retroelements by the three prime repair exonuclease 1 (TREX1) and excision repair cross complementing 1(ERCC) appear to mimic the action of RM-systems in bacteria, and the non-homologous end-joining (NHEJ) that follows the use of ZFN without a repair template.

Plant Transformation

Genetic transformation involves the integration of gene into genome by means other than fusion of gametes or somatic cells. The foreign gene (termed the "transgene") is incorporated into the host plant genome and stably inherited through future generations. This plant transformation approach is being used to generate plant processing trails, unachievable by conventional plant breeding, especially in case where there is no source of the desired trait in the gene pool. In the gene of interest, the correct regulatory sequences are incorporated i.e. promoters and terminators, and then the DNA is transferred to the plant cell or tissue using a suitable vector. The gene of interest is attached to a selectable marker which allows selection for the presence of the transgene. Confirmation for the presence of inserted genes is generally tested by resistance to a specific antibiotic present in the medium. Once the plant tissue has been transformed, the cells containing the transgene are selected and regeneration back into whole plants is carried out. This is possible as plant cells are totipotent, which means that they contain all the genetic sequence to control the development of that cell into a normal plant. Therefore, the gene of interest is present in every single plant cell; however, where its expression is controlled by the promoter. Plant transformation can be carried out by various ways depending on the species of the plant. A major method of DNA transfer in plants is *Agrobacterium* mediated transformation. *Agrobacterium is* a natural living soil bacteria and is capable of infecting a wide range of plant species, causing crown gall diseases. It has natural transformation abilities. When *A. tumefaciens* infects a plant cell, it transfers a copy of its T-DNA, which is a small section of DNA carried on its Ti (Tumour inducing) plasmid. This T-DNA is flanked by two (imperfect) 25 base pair repeats. Any DNA contained within these borders will be transferred to the host cell when used as transformation vector.

Different Types of Plant Transformation Vectors

Plant transformation vectors comprises of plasmids that have been purposely designed to facilitate the generation of genetically modified plants. The most commonly applicable plant transformation vectors are binary vectors which have the ability to replicate in *E. coli*, a common lab bacterium, as well as in *Agrobacterium tumefaciens*, bacterium used to insert the recombinant (customized) DNA into plants. Plant transformation vectors contain three essential elements:

- Plasmids selection (creating a custom circular strand of DNA)

- Plasmids replication (so that it can be easily worked with T-DNA)

- T-DNA region (inserting the DNA into the *Agrobacterium*)

Co-integrate pTi Vector

The discovery that the vir genes do not need to be in the same plasmid with a T-DNA region to lead its transfer and insertion into the plant genome led to the construction of a system for plant transformation where the T-DNA region and the vir region are on separate plasmids. A co-integrative vector produced by integration of recombinant intermediate vector (IV containing the DNA inserts) in to a disarmed pTi. Transformed gene is initially cloned in *E. coli* for easy in cloning procedure. A suitably modified *E. coli* plasmid is used to initiate cloning of gene. The subsequent gene transfer in to plants is obtained by co-integrative vectors. Co-integration of the two plasmids is achieved with in *Agrobacterium* by homologous recombination.

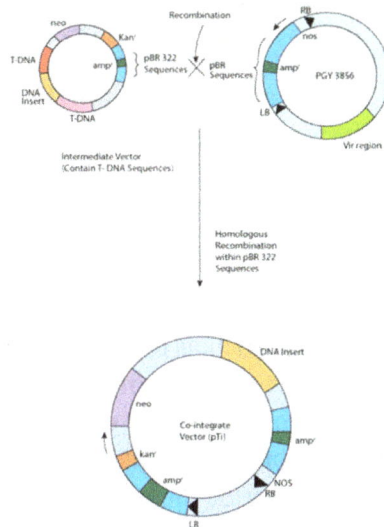

Diagrammatic representation of homologous recombination between disarmed pTi and recombinant IV (intermediate vector) containing the desired DNA insert to produce a cointegrative vector. (LB & RB – left and right borders of T-DNA; neo- neomycin phosphotransferase; kan r - Kanamycin resistance; ampr- ampicillin resistance).

Binary Vector

A binary vector consists of a pair of plasmids of which one contain *vir* region and other contains disarmed T-DNA sequence with right and left border sequences. The plasmid contain disarmed T-DNA are called micro-Ti or mini-Ti for e.g. Bin 19.

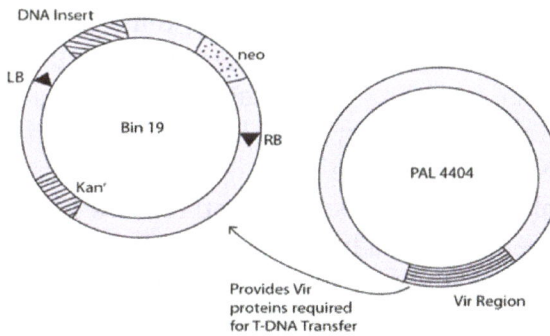

Binary vectors Bin19 and PAL 4404 of pTi

Plant Virus Vector

- Viruses have following features as a vector

- Infect cells of adult plant (dicotyledonous and monocotyledonous both)

- They produce large number of copies per cell which facilitate gene amplification and produce large quantities of recombinant protein.

• Some are systemic that they can spread throughout the plant.

Mostly plant viruses have RNA genome; two such viruses have great potential for vectors are brome mosaic virus (BMV) and tobacco mosaic virus (TMV). But maximum processes have been made with two DNA genome containing viruses as a vector, *viz* ., Caulimoviruses and Gemini viruses.

Cauliflower Mosaic Virus (CaMV)

The Cauliflower Mosaic Virus (CaMV) is a double-stranded DNA virus which infects a wide range of crucifers, especially Brassicas, such as cabbage, cauliflower, oilseed rape or mustard. In order to get itself and its DNA replicated (multiplied) within a plant cell, the virus must trick the plant's own molecular 'machinery' to do this task. For this purpose the virus has two promoters (35S and 19S) in front of its genes, which the plant cell believes to be its own. Furthermore, these promoters override the plant's own regulatory system, as they are constitutive, i.e. they are constantly switched on and can't be regulated or switched off by the plant. The CaMV 35S well known promoter is being used in almost all GM crops currently grown or tested, especially GM maize. It is the promoter of selection for plant genetic engineering, as it is a strong and constitutive promoter. Failure to distinguish or to ignore its capacity to be universally active in almost any organism is irresponsible and careless and shows a serious lack of scientific rigor and commitment to safety.

Gemini Viruses

Gemini viruses are small circular DNA viruses that replicate in plant nuclei. The Gemini virus vectors lack a coat protein gene, they are not transmissible by insect vectors, which are required for plant-to-plant spread and, thus, use of the disarmed vectors does not require a permit. Viruses from the Gemini virus family normally infects a wide range of crop plants, including maize, cotton, wheat, bean and cassava and are, therefore, an ideal system of choice for VIGS-based gene function analyses in a broad range of crop plants. Now vectors have been developed for use in cotton, and work is also ongoing for suitable vectors for roses. Using these new VIGS vectors, recombinant virus bearing a partial sequence of a host gene is used to infect the plant. As the virus spreads, the endogenous gene transcripts, which are homologous to the insert in the viral vector, are degraded by post-transcriptional gene silencing. These VIGS virus vectors have been used in a range of studies to silence single or multiple genes, including the meristematic gene, Proliferating Cell Nuclear Antigen (PCNA).

Tobacco Mosaic Virus (TMV)

TMV have single-stranded RNA genome which also serves as mRNA. It encodes at least four proteins in three open reading frames. Its genome contains 4 genes, of these the coat protein (cp) gene seems to be nonessential and can be site of integration of transgene. Viral RNA promoters are successfully manipulated for the synthesis of recombinant messenger

RNAs in whole plants. This vector consist of two steps, first, is the use of cDNA copy of viral genome for cloning in *E. coli* and, second, is *in vitro* transcription of the recombinant viral genome cDNA to produce infectious RNA copies to be used for plant infection.

Brome Mosaic Virus (BMV)

Brome mosaic virus (BMV) belongs to the family *Bromoviridae* of plant RNA viruses. BMV is a eukaryotic RNA virus, and its replication is entirely cytoplasmic. BMV genome is divided among three RNAs (1, 2 and 3) each packed into separate particle. Viral replication is dependent on well-organized interaction between nonstructural proteins 1a and 2a, encoded, respectively, by genomic RNA1 (gB1) and RNA2 (gB2). Genomic RNA3 (gB3) is dicistronic. Another nonstructural movement protein (MP) which promotes cell-to-cell spread encoded by 5′ half, while the capsid protein gene (CP) encoded in the 3′ half is translationally silent but is expressed from a subgenomic RNA (sgB4) that is synthesized from progeny minus-strand gB3 by internal initiation mechanisms. It was found in the absence of a functional replicase, assembled virions contained non-replicating viral RNAs (RNA1 or RNA2 or RNA3 or RNA1 + RNA3 or RNA2 + RNA3) as well as cellular RNAs. This indicates that placing a transgene downstream to the regulatory sequences of the *cp* gene of BMV will give high yields of the protein encoded by it.

Modes of Gene Delivery in Plant

Different systems are now available for gene transfer and successive regeneration of transgenic plants and the most common being *Agrobacterium* -mediated transformation. However, the preferred host of *Agrobacterium* is the dicot plants and its efficiency to transfer genes in monocots is still unsatisfactory. The alternative to this, is the introduction of DNA into plants cells without the involvement of a biological agent like, *Agrobacterium* , and leading to stable transformation is known as direct gene transfer. T he most often applied direct methods are microprojectile bombardment or protoplast transformation.

The direct DNA transfer methods have been subdivided into three categories:

1. Physical gene transfer method

2. Chemical gene transfer method

3. DNA imbibitions by cell, tissue and organ

Physical Gene Transfer method

Particle Bombardment

The Particle bombardment device, well known as the gene gun, was developed to enable penetration of the cell wall so that genetic material containing a gene of interest can be transferred into the cell. This physical direct gene transfer method, gene gun is used for genetic transformation of several organisms to introduce a diverse range of desirable traits. Plant transformation using particle bombardment follows the same steps as in *Agrobacterium* mediated transformation method:

i.   Isolation of desired genes from the source organism

ii. To develop a functional transgenic construct including the selected gene of interest; promoters to drive expression; modification of codon, if needed, to increase successful protein production; and marker genes to facilitate tracking of the introduced genes in the host plant

iii. Insertion of transgenic construct into a useful plasmid

iv. Introduce the transgenes into plant cells

v.   Regenerate the plants cells, and

vi. Test the performance of traits or gene expression under *in vitro*, greenhouse and field conditions.

A gene gun apparatus

In particle bombardment method, 1-2 µm tungsten or gold particles (called micro-projectiles) coated with genetically engineered DNA are accelerated with air pressure at high velocities and shot into plant tissues on a Petri-plate, as shown in figure. This is the second most widely used method, after *Agrobacterium* mediated transformation, for plant genetic transformation. The device accelerates particles in one of the two ways: (1) by means of pressurized helium gas or (2) by the electrostatic energy released by a droplet of water exposed to high voltage. The earlier devices used blank cartridges in a modified firing mechanism to provide the energy for particle acceleration, and thus, the name particle gun. It is also called Biolistics, Ballistics or Bioblaster).

The microcarriers (or microprojectiles), the tungsten or gold particles coated with DNA, are carried by macrocarriers (macro projectiles) which are then inserted into the apparatus and pushed downward at high velocities. The Macro-projectile is stopped by a perforated plate, while allowing the microprojectiles to propelled at a high speed into the plant cells on the other side. As the micro-projectiles enter the plant cells, the transgenes are free from the particle surface and may inserted into the chromosomal DNA of the plant cells. Selectable markers help in identifying those cells that take up the transgene or are transformed. The transformed plant cells are then regenerated and developed into whole plants by using tissue culture technique.

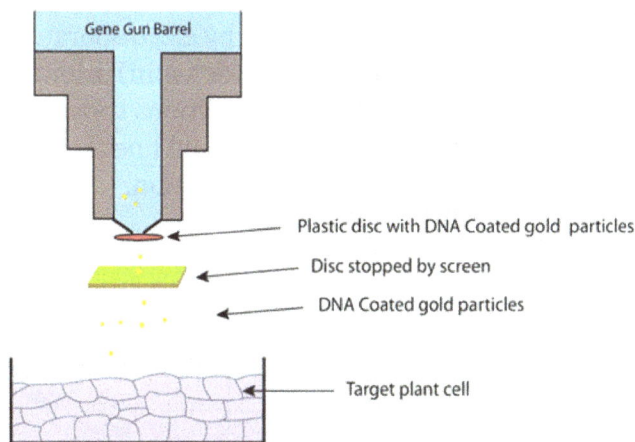

Diagrammatic illustration of gene transfer using Gene Gun method

The technique has many advantages and can be used to deliver DNA into virtually all the tissues, like immature and mature embryos, shoot-apical meristem, leaves, roots etc. Particle bombardment methods are also useful in the transformation of organelles, such as chloroplasts, which enables engineering of organelle-encoded herbicide or pesticide resistance in crop plants and to study photosynthetic processes.

Limitations to the particle bombardment method, compared to *Agrobacterium*-mediated transformation, include frequent incorporation of multiple copies of the transgene at a single insertion site, rearrangement of the inserted genes, and insertion of the transgene at multiple insertion sites. These multiple copies can be associated with silencing of the transgene in subsequent progeny. The target tissue may often get damaged due to lack of control of bombardment velocity.

Electroporation

Electroporation is another popular physical method for introducing new genes directly into the protoplasts. In this method, electric field is playing important role. Due to the electric field protoplast get temporarily permeable to DNA. In electroporation, plant cell protoplasts are kept in an ionic solution containing the vector DNA in a small chamber that has electrodes at opposite ends. A pulse of high voltage is applied to the electrode which makes the transient pores (ca. 30 nm) in the plasma membrane, allowing the DNA to diffuse into the cell. Immediately, the membrane reseals. If appropriately treated, the cells can regenerate cell wall, divide to form callus and, finally, regenerate complete plants in suitable medium. The critical part of the procedure is to determine conditions which produce pores that are sufficiently large and remain open long enough to allow for DNA diffusion. At the same time, the conditions should make pores that are temporary. With a 1 cm gap between the electrodes and protoplasts of 40-44µm diameter, 1-1.5 kVcm^{-2} of field strength for 10µs is required for efficient introduction of DNA. It was seen that presence of 13% PEG (added after DNA) during electroporation significantly raised the transformation frequency. The other factors which may improve the transformation frequency by electroporation are linearizing of plasmid, use of carrier DNA, and heat shock (45 ~ for 5 min) prior to addition of vector, and placing on ice after pulsing. Under optimal conditions transformation frequencies of up to 2% have been reported. Stably transformed cell lines and full plants of a number of cereals have been produced through electroporation.

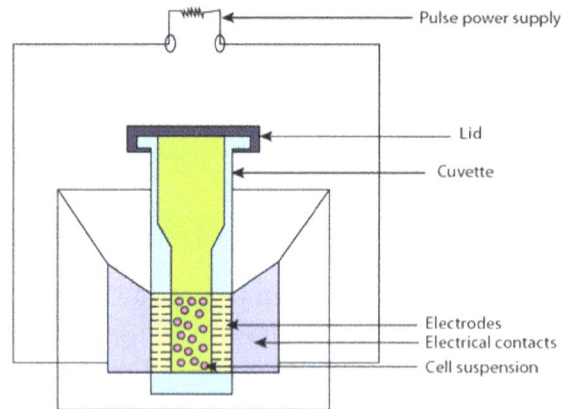

Electroporation

There are some parameters that can be considered when performing *in vitro* electroporation:

1. Cell size

Cell size is inversely correlated to the size of the external field needed to generate permeabilization. Consequently, optimization for each cell type is essential. Likewise, cell orientation matters for cells that are not spherical.

2. Temperature

It has been observed that plant membrane resealing is effectively temperature dependent and shows slow closure at low temperatures. For DNA transfer, it has been found that cooling at the time of permeabilization and subsequent heating in incubator increases transfer efficacy and cell viability.

3. Post-pulse manipulation

Cells are susceptible when in the permeabilized state, and it has been shown that waiting for 15min after electroporation in order to allow resealing before pipetting cells, increases cell viability.

4. Composition of electrodes and pulsing medium

For short pulses is needed for release of metal from the standard aluminium electrodes used in standard disposable cuvettes. Some authors advocate the use of low conductivity or more resistance media for DNA transfer in order to increase viability and increase transfection efficacy.

Microinjection

The microinjection technique is a direct physical approach to inject DNA directly into the plant protoplasts or cells (specifically into the nucleus or cytoplasm) using fine tipped (0.5-1.0 μm diameter) capillary glass needle or micropipettes. Through microinjection technique, the desired gene introduce into large cells, such as oocytes, eggs, and the cells of early embryo.

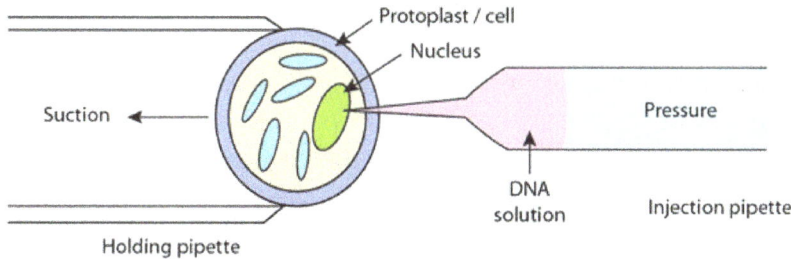

Microinjection

Liposome-mediated Transformation

The idea of a method of direct plant transformation elaborated in the middle eighties was to introduce DNA into the cell by means of liposomes. Liposomes are microscopic spherical vesicles that form when phospholipids are hydrated. Liposomes are circular lipid molecules with an aqueous interior that can carry nucleic acids. Liposomes encapsulate the DNA fragments and then adhere to the cell membranes and fuse with them to transfer DNA fragments. Thus, the DNA enters the cell and then to the nucleus. Lipofection is a very efficient technique used to transfer genes in bacterial, animal and plant cells. They can be loaded with a great variety of molecules, including DNA. In the case of protoplasts, the transfection (lipofection) occurs through the membrane fusion and endocytosis. When pollen grains are transformed, liposomes are delivered inside through pores. The efficiency of bioactive-beads-mediated plant transformation was improved using DNA-lipofection complex as the entrapped genetic material instead of naked DNA used in the conventional method. Liposome-mediated transformation is far from routine, in spite of the low expense and equipment requirement. A probable reason is its laboriousness and low efficiency. Only few reports on the integration of genes introduced by means of liposomes followed by transgenic plant regeneration for tobacco and wheat have been published thus far.

Silicon Carbide Fiber Mediated Transformation (SCMT)

SCMT is one of the least complicated methods of plant transformation. Silicon carbide fibers are simply added to a suspension containing plant tissue (cell clusters, immature embryos, callus) and plasmid DNA, and then mixed in a vortex, or in other laboratory apparatus such as, commercial shakers, blenders etc. DNA-coated fibers penetrate the cell wall in the presence of small holes created in collisions between the plant cells and fibers. The most often used fibers in this procedure are single crystals of silica organic minerals like, siliconcarbide, which have an elongated shape, a length of 10–80 mm, and a diameter of 0.6 mm, and which show a high resistance to expandability. Fiber size, the parameters of vortexing, the shape of the vessels used, the plant material and the characteristics of the plant cells, especially the thickness of the cell wall are the factors depending on the efficiency of SCMT. There are several known examples of deriving transgenic forms, cell colonies or plants in maize, rice, Wheat, tobacco, *Lolium multiflorum*, *Lolium perenne, Festuca arundinacea*, *and Agrostis stolonifera* by SCMT.

SCMT is an easy, fast and inexpensive procedure. Therefore, it could be an attractive alternative method of plant transformation in particular situations, e.g. when a gene gun is not available and *Agrobacterium* -mediated transformation is difficult or not possible (as in the case of numerous monocots). The other advantages of the SCF-mediated method over other procedures include the ability to transform walled cells, thus, avoiding protoplast isolation .

The main disadvantages of this method are low transformation efficiency, damage to cells, thus, negatively influencing their further regeneration capability. Another disadvantage is that silicon fibers have similar properties to asbestos fibers and care must be taken when working with them as breathing the fibers can lead to serious sicknesses. Silicon carbide has some carcinogenic properties as well.

The Pollen-tube Pathway Method

The transformation method via pollen-tube pathway has great function in agriculture molecular breeding. Foreign DNA can be applied to cut styles shortly after pollination. The DNA reaches the ovule by flowing down the pollen-tube. This procedure, the so-called pollen-tube pathway (PTP), was applied first time for the transformation of rice. The authors obtained transgenic plants at remarkably high frequency. Afterward PTP was used for other species e.g. wheat, soybean, *Petunia hybrida* and watermelon. A bacterial inoculum or plasmid DNA can also be injected into inflorescence with pollen mother cells in the pre- meiotic stage without removing the stigma. In that case, it is expected that foreign DNA will be integrated with the gamete genome. Such an approach has been employed for rye. Pollen collected from inflorescences injected with a suspension of genetically engineered *A. tumefaciens* strain was predestined for the pollination of the emasculated spikes of the maternal plant. But the transformation efficiency was about 10-fold lower than that approximately reached for this species via microprojectile bombardment. Shou et al. (2002) also reported they were unable to reproduce the pollen-tube pathway transformation for delivering plasmid DNA into soybean. They concluded that the pollen-tube pathway transformation in cotton and soybean was not reproducible. This might have been because of the manipulation of transformation, the growth stage of plants, the effects of environment and weather

Chemical Gene Transfer Method

This involves plasma membrane destabilizing and/or precipitating agents. Protoplasts are mainly used which are incubated with DNA in buffers containing PEG, poly L-ornithine, polyvinyl alcohol or divalent ions. The chemical transformation techniques work for a broadspectrum of plants.

Polybrene–Spermidine Treatment

The combination polybrene–spermidine treatment greatly enhanced the uptake and expression of DNA and, hence, the recovery of nonchimeric germline transgenic cotton plants. The major advantages of using the polybrene–spermidine treatment for plant genetic transformation are that polybrene is less toxic than the other polycations; spermidine protects DNA from shearing because of its condensation effect; and because no carrier DNA is used, and the integration of plasmid DNA into the host genome should enable direct analysis of the sequences surrounding the site of integration. To deliver plasmid DNA into cotton suspension culture obtained from cotyledon-induced callus, polybrene and/or spermidine treatments were used. The transforming plasmid (pBI221.23) contained the selectable hpt gene for hygromycin resistance and the screenable gus gene. Primary transformant cotton plants were regenerated and analyzed by DNA hybridization and b-glucuronidase assay.

PEG Mediated Gene Transfer

In this method protoplasts are isolated and a particular concentration of protoplast suspension is taken in a tube followed by addition of plasmid DNA (donor or carrier). To this 40% PEG4000(w/v) dissolved in mannitol and calcium nitrate solution is slowly added because of high viscosity, and this mixture is incubated for few minutes (ca 5 min.). As per the requirements of the experiments, transient or stable transformation studies are conducted. Among the most important parameters that affect the efficiency of PEG-mediated gene transfer are the concentration of calcium and magnesium ions in the incubation mixture, and the presence of carrier DNA. The linearized dsDNA are more efficiently expressed and integrated in the genome than the supercoiled forms. The advantage of the method is that the form of DNA applied to the protoplast is controlled entirely by the experimenter and not by intermediate biological vector. Main disadvantage is that the system requires a protoplast.

Calcium-Phosphate Co-precipitation

DNA when mixed with calcium chloride solution isotonic phosphate buffer DNA-CaPO$_4$ precipitate. The precipitate is allowed to react with actively dividing cells for several hours, washed and then incubated in the fresh medium. Giving them a physiological shock with DMSO can increase the efficiency of transformation to a certain extent. Relative success depends on high DNA concentration and its apparent protection in the precipitate.

DEAE Dextran Procedure

Transformation of cells with DNA complexed to the high molecular weight diethyl amino ethyl (DEAE) dextran is used to obtain efficient transient expression. The efficiency increasewhen 80% DMSO shock is given. But this technique does not produce stable transformants.

The Polycation DMSO Technique

It involves use of a polycation, polybrene, to increase the absorption of DNA to the surface followed by a brief treatment by 25-30% DMSO to increase the membrane permeability and enhance the uptake. The major advantage of polybrene is that it is less toxic than other polycations and a high transformation efficiency requires very small quantities of plasmid DNA to be used.

Direct Gene Transformation Through Imbibition

During imbibition the uptake of exogenous DNA of dehydrated plant tissues is a direct gene transfer method which has been studied since the 1960s and for which the literature contains a number of both claims and refutations. The physical and biochemical changes which are already known occur in plant tissues during dehydration (e.g. a large water potential between the dry tissue and external solution, rapid cell expansion, cell wall rupture, cell membrane structural changes and leakiness; suggest that under these conditions DNA uptake might be possible. DNA uptake and expression was observed under simple dehydration conditions, but was stimulated by the presence of 20% DMSO, suggesting that membrane permeabity was an important factor in the process. A number of lines of evidence supported the conclusion that reporter gene expression was the result DNA uptake into cells and plants were recovered from treated embryos, but no evidence of stable transformation was presented. Subsequent research on the imbibition transformation has extended its application to dessicated somatic embryos of alfalfa, which showed transient GUS expression at frequencies upto 70%. The stable transformation of rice by embryo imbibition was also reported. The frequency of transient expression of gusA and hpt genes using the CaMV35S promoter was about 30 to 50%. The main sites of gusA gene expression were meristems of roots and vascular bundles of leaves. Also, DNA uptake, integration and expression of the hpt gene in selected rice were investigated by various PCR methods and Southern blot analysis of genomic DNA. It was shown that the hygromycin phosphotransferase (HPT) DNA was present in the rice genome in an integrated form and not as a plasmid form. These methods are technically the most simple of DGT methods, as they require no specialist equipment and the preparation of target plant tissues are generally simple. This simplicity constitutes the advantage of these techniques, while their limitations are i) they can be applied only to very specific organs or tissues (i.e. newly pollinated flowers or hydrating embryos) and ii) it is still not clear that they lead to stable, and heritable transformation. While they add support to the observation that many different plant cells may be amenable to DNA uptake and expression, at present these techniques are subjects to further analysis and development rather than usable gene transfer methods.

Agrobacterium Mediated Gene Transfer

Agrobacterium tumefaciens and *Agrobacterium rhizogenes* are common gram-negative soil borne bacteria causing induction of 'crown gall' and 'hairy root' diseases. These bacteria naturally insert their genes into the genome of higher plants. The studies on crown gall formation revealed that the virulent strains of bacteria introduce a part of their genetic material into the infected cells where it gets integrated randomly with the genetic material of the host cell. The bacterial genes are able to replicate along with the plant genome and uses the machinery of plants to express their genes in terms of the synthesis of a special class of compounds, called opines, which the bacterium uses as nutrients for its growth but are useless to the host cells. In the process, *Agrobacterium* causes plant tumors (gall formation) commonly seen near the junction of the root and the stem and is called 'crown gall disease'. *A. tumefaciens* attracted to the wound site via chemotaxis, in response to chemicals (sugars and phenolic molecules) released from the damaged plant cells. The disease afflicts a great range of dicotyledonous plants, which constitute one of the major groups of flowering plants. Tumorous plant cells were found to contain DNA of bacterial origin integrated in their genome. Furthermore, the transferred DNA (named T-DNA) was originally part of a small molecule of DNA located outside the chromosome of the bacterium. This DNA molecule was called Ti (tumor-inducing) plasmid.

Plant Gene Structure

Plant ribosomal RNA genes and a number of other structural genes from a variety of species have now been analyzed in considerable detail. In common with many ani-mal genes, some plant gene sequences have been found to have their coding sequenc-es interrupted by introns or intervening sequences. These introns are transcribed but not represented in mature mRNA and hence, are not translated. No introns have been found in rRNA genes but they have been demonstrated in a number of other plant structural genes. A typical plant gene is shown in figure.

A typical plant gene has the following region beginning with the 5'end:

   i). Promoter: For transcription initiation

   ii). Enhancer/silencer: Concerned with regulation of gene

   iii). Transcriptional start site or cap site: From here initiation of transcription take place

   iv). Leader sequence: It is untranslated region

   v). Initiation codon

   vi). Exons

   vii). Introns

   viii). The stop codon

   ix). A second untranslated region, and

   x). Poly A tail

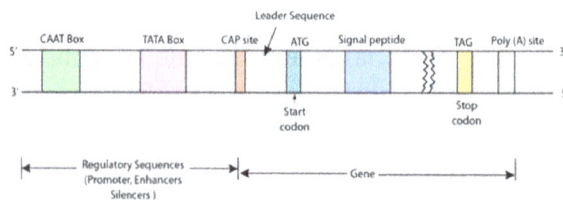

A typical plant gene

Promoter is a region of DNA sequence which helps in the transcription of a particular gene. This contains specific DNA sequences as well as response elements which provide a secure initial binding site for RNA polymerase. These proteins called transcription factors that recruit RNA polymerase. The CAAT and TATA boxes represent consensus sequences within promoter for RNA polymerase II. ATG (AUG in mRNA) is initiation codon for mRNA translation, and mark the beginning of coding sequence of the gene. A sequence between the cap site and ATG is not translated and form the 5'-leader se-quence of mRNA. Codon TAG/TAA/TGA are chain terminating codon and it is fol-lowed by a stretch of nontranslated region. At the end, poly-adenylation site is present which denotes the end of transcription.

Organization of T-DNA

The transfer DNA (T-DNA) is the transferred DNA of the tumour inducing plasmid (pTi) of some *Agrobacterium* species of bacteria. T-DNA has both its side 24 kb direct repeat border sequence and contains the gene for tumor / hairy root induction and also for opines biosynthesis. pTi has three genes, two of these genes (*iaaM* and *iaaH*) encode enzymes which together convert tryptophane in to IAA (Indol-3-acetic acid) a type of auxin. If these two genes are deleted then shooty crown gall will produce. Therefore, the locus was earlier called 'shooty locus' and the genes were designated as *tms* 1 (tumour with shoots) and *tms* 2. The third gene, *ipt*, encodes an enzyme which produces Zeatin-type cytokinin isopentenyl adenine. The deletion of *ipt*, causes rooty crown galls and the region was earlier designated as 'rooty locus' and denoted by *tmr* (tumour having roots). In addition to these, another locus called *tml* and the deletion of which results in large tumours. Besides, T-DNA also contains genes involved in opine biosysnthesis which are located near the right border of T-DNA.

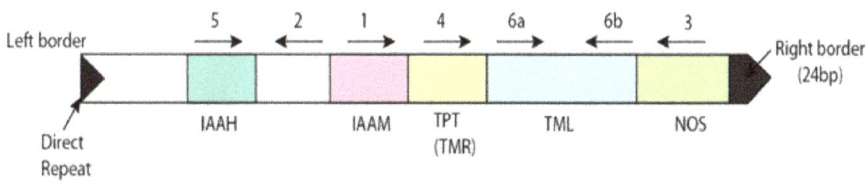

Nopaline type Ti plasmid T-DNA (Arrows indicating the direction of transcription and number indicates the transcriptional unit)

T-DNA Transfer and Integration

The steps involved in T-DNA transfer and integration in to the plant genome are explained in figure.

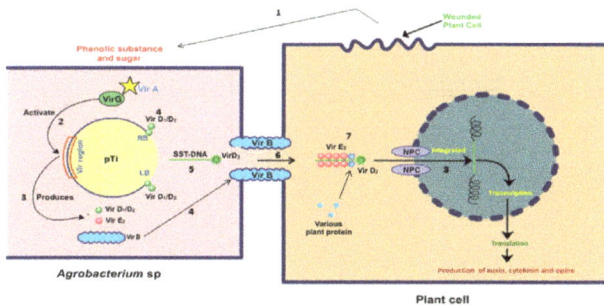

T-DNA transfer and integration

Wounded plant cell releases phenolics substances and sugars (1); which are sensed by *vir* A, *vir* A activates *vir* G, *vir* G induces expression of *vir* gene of Ti-plasmid (2); *vir* gene produce all the *vir* -protein (3); *vir* D_1 and *vir* D_2 are involve in ssT-DNA production from Ti-plasmid and its export (4) and (5); the ssT-DNA (with associated *vir* D_1

and vir D_2) with vir E_2 are exported through transfer apparatus vir B (6); in plant cell, T-DNA coated with vir E_2 (7); various plant proteins influence the transfer of T-DNA + vir D_1 + vir D_2 + vir E_2 complex and integration of T-DNA to plant nuclear DNA(8). (LB= left border; RB= Right border; pTi = Ti plasmid, NPC = nuclear pore complex)

Signal Recognition by *Agrobacterium* spp.

The wounded plant cells release certain chemicals, such as phenolics and sugars. These chemicals are recognized by *Agrobacterium* as signals. This in turn results in a sequence of biochemical events in *Agrobacterium* that helps in transfer of T-DNA of Ti plasmid.

Attachment to Plant Cell

Attachment of this bacterium to plant cells is a two step process. It involves an initial attachment via a polysaccharides (the product of *att* R locus). Subsequently, a mesh of cellulose fibres is produced by *Agrobacterium*. Several chromosomal virulence genes (*chv* genes) are involved in attachment of bacterial cells to the plant cells.

Induction of Virulence Gene

vir A (a membrane-linked sensor kinase) senses phenolics (such as acetosyringone) and autophosphorylates, subsequently phosphorylating and, thereby, activating vir G. This activated vir G induces expression of virulence gene of Ti plasmid to produce the corresponding virulence proteins (D, D2, E2, B). It has been also identified that certain sugars (e.g. glucose, galactose, xylose etc.) also induce virulence gene.

Table: *Agrobacterium* virulence protein function

Virulence protein	Function in *Agrobacterium* spp.	Function in plant
virA	• Phenolic sensor • Part of two component system with VirG phosphorylation and activates VirG	-
virG	• Transcriptional factor • Responsible for *vir* gene expression	-
virB1-B11	Components of membrane structure for T-DNA transfer	-
virD1	• In T-DNA processing • Modulate *vir*-D2 activity	-
virD2	• Nick the T-DNA • Directs the T-DNA through *vir*B transfer apparatus	-
virE2		• Single stranded DNA-binding protein • Prevents T-DNA degradation by nucleases • Involved in nuclear targeting and helps in passage through nuclear pore complex (NPC)

Production of T-DNA Strand

The right and left border sequence of T-DNA are identified by vir D1/ vir D2 protein complex and vir D2 produces single stranded DNA (ss-T-DNA). After nicking, vir D2 becomes covalently attached to the 5'end of ss-T- DNA strand and protect and export the ss-T-DNA to plant cells.

Transfer of T-DNA out the Bacterial Cell

The ss-T-DNA – *vir* D2 complex in association with *vir* E2 is exported from bacterial cell by a 'T-pilus' (a membrane channel secretary system).

Transfer T-DNA into Plant Cell and Integration

The single stranded T-DNA–*vir* D2 complex and other *vir* proteins cross the plant plasma membrane. In the plant cells, T-DNA gets covered with *vir* E2. This covering of *Vir* E2 helps in protection of ss-T-DNA from degradation by nucleases. *vir* D2 and *vir* E2 interact with variety of plant proteins which influence the T-DNA transport and integration. The T-DNA – *Vir* D2 – *Vir* E2 – plant proteins complex enters the nucleus through nuclear pore complex (NPC). In the nucleus, T-DNA gets integrated into the plant genome by a process referred to as 'illegitimate recombination'. This process is unlike homologous recombination as it does not depend on extensive region of sequence similarity.

Ti and Ri Plasmids

Agrobacterium species harboring tumor-inducing (Ti) or hairy root-inducing (Ri) plasmids cause crown gall or hairy root diseases, respectively in plants. *Agrobacterium tumefaciens* is a plant pathogen that induces tumor on a wide variety of dicotyledonous plants and the disease is caused by tumor-inducing plasmid (pTi). Similarly *Agrobacterium rhizogenes* is a plant pathogen that induces hairy roots on a wide variety of dicotyledonous plants and the disease is caused by root-inducing plasmid (pRi). Virulence *(vir)* genes of Ri as well as of Ti plasmids are essential for the T-DNA transfer into plant chromosomes . These natural plasmids provide the basis for vectors to make transgenic plants. The plasmids are approximately 200 kbp in size. Both pTi and pRi are unique in two respects: (i) they contain some genes, located within their T-DNA, which have regulatory sequences recognized by plant cells, while their remaining genes have prokaryotic regulatory sequences, (ii) both plasmids naturally transfer a part of their DNA, the T-DNA, into the host genome, which makes *Agrobacterium* a natural genetic engineer.

Complete sequence analysis confirms that the pathogenic plasmids contain gene clusters for DNA replication, virulence, T-DNA, opine utilization and conjugation. T-DNA genes have lower G + C content, which is presumably suitable for expression in host plant cells. Besides these genes, each plasmid has a large number of unique genes. Even plasmids of the same opine type differ considerably in gene content and are highly chimeric in structures. The plasmids seem to interact with each other and with plasmids of other members of the *Rhizobiaceae* and are likely to shuffle genes of infection between Ti and Ri plasmids. Plasmid stability genes are talked about, which are important for plasmid evolution and construction of useful strains.

Ti Plasmid

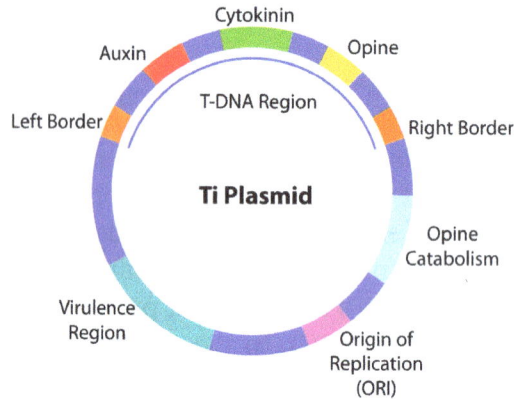

The structure of the Ti plasmid

A Ti or tumour inducingplasmid is a plasmid that often, but not always, is a part of the genetic equipment that *Agrobacterium tumefaciens* and *Agrobacterium rhizogenes* use to transduce its genetic material to plants. The Ti plasmid is lost when *Agrobacterium* is grown above 28 °C. Such cured bacteria do not induce crown galls, i.e. they become avirulent. pTi and pRi share little sequence homology but are functionally rather similar. The Ti plasmids are classified into different types based on the type of opine produced by their genes. The different opines specified by pTi are octopine, nopaline, succinamopine and leucinopine.

The plasmid has 196 genes that code for 195 proteins. There is one structural RNA. The plasmid is 206,479 nucleotides long, the GC content is 56% and 81% of the material is coding genes. There are no pseudogenes.

The modification of this plasmid is very important in the creation of transgenic plants.

Virulence Region

Genes in the virulence region are grouped into the operons*virABCDEFG*, which code for the enzymes responsible for mediating conjugative transfer of T-DNA to plant cells.

- *virA* codes for a receptor which reacts to the presence of phenolic compounds such as acetosyringone,syringealdehyde or acetovanillone which leak out of damaged plant tissues.

- *virB* encodes proteins which produce a pore/pilus-like structure.

- *virC* binds the overdrive sequence.

- *virD1* and *virD2* produce endonucleases which target the direct repeat borders of the T-DNA segment; virD4 is the coupling protein.

- *virE* binds to T-strand protecting it from nuclease attack, and intercalates with lipids to form channels in the plant membranes through which the T-complex passes, beginning with the right border.

- *virG* activates vir-gene expression after binding to a consensus sequence, once it has been phosphorylated by *virA*.

Characteristic Features

- *Agrobacterium* is called the natural genetic engineer.

- Size of the plasmid: ~250 kbp.

- Contains one or more T-DNA region.

- Contains a region enabling conjugative transfer.

- Contains regions for opine synthesis and catabolism.

- Responsible for crown gall disease in plants.

Ri Plasmid

Agrobacterium rhizogenes is a soil born gram negative bacterium. It causes hairy root disease of many dicotyledonous plants. The ability of *A. rhizogenes* to incite hairy root disease is confirmed by a virulence plasmid, which is similar to that found in *Agrobacterium tumefaciens* which causes Crown gall tumors of plants. The virulence plasmid of *A. rhizogenes* is commonly known as the Ri-plasmid (pRi). The pRi have extensive functional homology with the pTi. The pRi contains distinct segment(s) of DNA, which is transferred to plant genome during infection. The transfer T-DNA to the plant genome is mediated by another segment on the plasmid known as the virulence (vir) region. All strains of *A. rhizogenes* are known to produce agrocinopine.

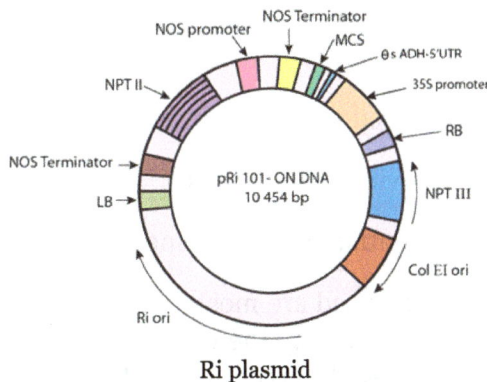

NOS Terminator
NOS promoter
MCS
θs ADH-5'UTR
NPT II
35S promoter
NOS Terminator
RB
LB
pRi 101- ON DNA
10 454 bp
NPT III
Col EI ori
Ri ori

Ri plasmid

Genetic Selection and Screening of Transformations

Genetic selection of transformed cells is a significant step of any plant transformation. Screening of transformed cells or plants for gene integration and expression in transformed cells or plants is a process that involves several techniques, including DNA and RNA blot hybridization analysis, PCR, ELISA analysis. In the absence of a correct selection system one would face with the option of screening every shoot that regenerates in a transformation experiment. In cases where transformation frequency is high this may be possible but for plant species that transform with low frequencies this would be a laborious, if not impossible, task. Therefore, a selectable marker gene is incorporated into the plant transformation vectors and an appropriate selecting agent is added to the culture medium which favors the growth of only transformed cells. The genes used as selectable markers are dominant and typically of bacterial origin. For successful selection, the target plant cells must be susceptible to moderately low concentrations of the selecting agent in a non-leaky way. The compound that inhibits the growth but does not kill the wild type cells is preferred as a selecting agent in plant transformation. The concentration of the selecting agent used varies widely depending on the sensitivity of the plant species and/or explant source.

Table: Selectable marker genes used in plant transformation

Gene	Enzyme encoded	Selective agent(s)
Antibiotics		
ble	Enzymic activity not known	Bleomycin
dhfr	Dihydrofolate reductase	Methotrexate Trimethoprim
hpt	Hygromycin phosphotransferase	Hygromycin B
npt II	Neomycin phosphotransferase	G418 Kanamycin
Herbicides		
als	Mutant forms of acetolactate synthase	Chlorsulfuron Imidazolinones
Aro A	5-Enolpyruvylshikimate 3-phosphate synthase	Glyphosate (Roundup)
bar	Phosphinothricin acetyltransferase	Phosphinothricin (Bialaphos)

A screening can also be possible by screening or scorable or reporter gene, incorporated into the transformation vectors, which allows for the detection of transformed cells, tissues or plants. The essential features of an ideal reporter gene are:

i.   An efficient and easy detection with high sensitivity

ii. Lack of endogenous activity in plant cells

iii. A relatively rapid degradation of the enzyme

The screening markers presently used are mostly derived from bacterial genes coding for an enzyme that is readily detected by the use of chromogenic, fluorigenic, photon emitting or radioactive substrates. A screening marker gene is functional only if an enzyme with comparable activity is not present in non-transformed cells. The utility of any particular gene construct as a transformation marker varies depending on the plant species and the tissue involved. The kanamycin resistance gene is probably the most extensively used selectable marker phenotype and *Uid* A gene (also referred to as *gus*), which encodes β-glucuronidase, is the most versatile reporter gene. The screened cells and the plants regenerated from transformation are further subjected to biochemical analyses, such as Southern hybridization, PCR and Northern hybridization. The former determines the presence and the number of copies of the introduced gene while the latter demonstrates the presence of transcripts of the transgene.

Table: Screenable marker genes used in plant transformation

Gene	Enzyme encoded	Substrate(s) and assays
CAT	Chloramphenicol acetyl transferase	[^{14}C]chloramphenicol and acetyl CoA; TLC separation of acetylated [^{14}C]chloramphenicol - detection by autoradiography
lac Z	β-galactosidase	As β-glucuronidase; problems with background activity in some species
GUS	β-glucuronidase	Range of substrates depending on assay; colourimetric, fluorometric, and histochemical techniques available
lux	Luciferase: bacterial insect	Decanal and FMNH$_2$ ATP and O$_2$ and luciferin Bioluminescent assays: quantitative tests on extracts or in situ tissue assays with activity detected by exposure of X-ray film
npt-II	Neomycin phosphoryltransferase	Kanamycin and [^{32}P]ATP In situ assay on enzyme fractionated by non-denaturing PAGE; enzyme detected by autoradiography Quantitative dot-binding assay on reaction products

Polymerase Chain Reaction

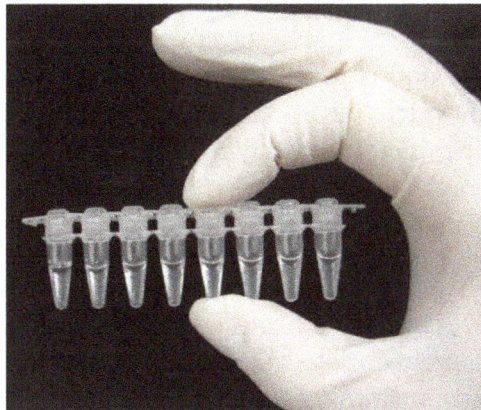

A strip of eight PCR tubes, each containing a 100 µl reaction mixture

Polymerase chain reaction (PCR) is a technique used in molecular biology to amplify a single copy or a few copies of a segment of DNA across several orders of magnitude, generating thousands to millions of copies of a particular DNA sequence. It is an easy, cheap, and reliable way to repeatedly replicate a focused segment of DNA, a concept which is applicable to numerous fields in modern biology and related sciences.

Developed in 1983 by Kary Mullis, PCR is now a common and often indispensable technique used in clinical and research laboratories for a broad variety of applications. These include DNA cloning for sequencing, construction of DNA-based phylogenies, or functional analysis of genes; diagnosis and monitoring of hereditary diseases; analysis of genetic fingerprints for DNA profiling (for example, in forensic science and parentage testing); and detection of pathogens in nucleic acid tests for the diagnosis of infectious diseases. In 1993, Mullis was awarded the Nobel Prize in Chemistry along with Michael Smith for his work on PCR.

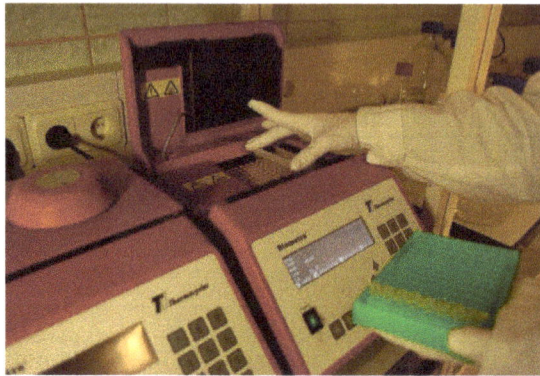
Placing a strip of eight PCR tubes into a thermal cycler

The vast majority of PCR methods rely on thermal cycling, which involves exposing the reactants to cycles of repeated heating and cooling, permitting different temperature-dependent reactions—specifically, DNA melting and enzyme-driven DNA replication—to quickly proceed many times in sequence. Primers (short DNA fragments) containing sequences complementary to the target region, along with a DNA polymerase, after which the method is named, enable selective and repeated amplification. As PCR progresses, the DNA generated is itself used as a template for replication, setting in motion a chain reaction in which the original DNA template is exponentially amplified. The simplicity of the basic principle underlying PCR means it can be extensively modified to perform a wide array of genetic manipulations. PCR is not generally considered to be a recombinant DNA method, as it does not involve cutting and pasting DNA, only amplification of existing sequences.

Almost all PCR applications employ a heat-stable DNA polymerase, such as Taq polymerase, an enzyme originally isolated from the thermophilic bacterium *Thermus aquaticus*. This DNA polymerase enzymatically assembles a new DNA strand from free nucleotides, the building blocks of DNA, by using single-stranded DNA as a template and DNA oligonucleotides (the primers mentioned above) to initiate DNA synthesis.

In the first step, the two strands of the DNA double helix are physically separated at a high temperature in a process called DNA melting. In the second step, the temperature is lowered and the two DNA strands become templates for DNA polymerase to selectively amplify the target DNA. The selectivity of PCR results from the use of primers that are complementary to the DNA region targeted for amplification under specific thermal cycling conditions.

Principles

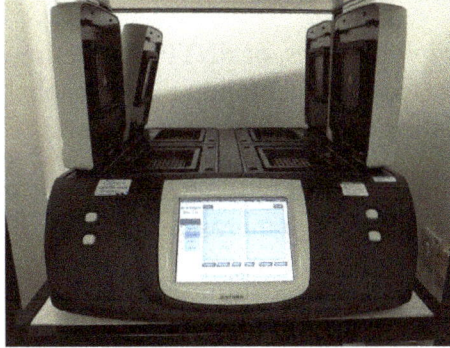

A thermal cycler for PCR

An older model three-temperature thermal cycler for PCR

PCR amplifies a specific region of a DNA strand (the DNA target). Most PCR methods amplify DNA fragments of between 0.1 and 10 kilo base pairs (kbp), although some techniques allow for amplification of fragments up to 40 kbp in size. The amount of amplified product is determined by the available substrates in the reaction, which become limiting as the reaction progresses.

A basic PCR set-up requires several components and reagents, including:

- a *DNA template* that contains the DNA target region to amplify

- a *DNA polymerase*, an enzyme that polymerizes new DNA strands; heat-resistant Taq polymerase is especially common, as it is more likely to remain intact during the high-temperature DNA denaturation process

- two DNA *primers* that are complementary to the 3' (three prime) ends of each of the sense and anti-sense strands of the DNA target (DNA polymerase can only bind to and elongate from a double-stranded region of DNA; without primers there is no double-stranded initiation site at which the polymerase can bind); specific primers that are complementary to the DNA target region are selected beforehand, and are often custom-made in a laboratory or purchased from commercial biochemical suppliers

- *deoxynucleoside triphosphates*, or dNTPs (sometimes called "deoxynucleotide triphosphates"; nucleotides containing triphosphate groups), the building blocks from which the DNA polymerase synthesizes a new DNA strand

- a *buffer solution* providing a suitable chemical environment for optimum activity and stability of the DNA polymerase

- *bivalentcations*, typically magnesium (Mg) or manganese (Mn) ions; Mg^{2+} is the most common, but Mn^{2+} can be used for PCR-mediated DNA mutagenesis, as a higher Mn^{2+} concentration increases the error rate during DNA synthesis

- *monovalent cations*, typically potassium (K) ions

The reaction is commonly carried out in a volume of 10–200 μl in small reaction tubes (0.2–0.5 ml volumes) in a thermal cycler. The thermal cycler heats and cools the reaction tubes to achieve the temperatures required at each step of the reaction. Many modern thermal cyclers make use of the Peltier effect, which permits both heating and cooling of the block holding the PCR tubes simply by reversing the electric current. Thin-walled reaction tubes permit favorable thermal conductivity to allow for rapid thermal equilibration. Most thermal cyclers have heated lids to prevent condensation at the top of the reaction tube. Older thermal cyclers lacking a heated lid require a layer of oil on top of the reaction mixture or a ball of wax inside the tube.

Procedure

Typically, PCR consists of a series of 20–40 repeated temperature changes, called cycles, with each cycle commonly consisting of two or three discrete temperature steps. The cycling is often preceded by a single temperature step at a very high temperature (>90 °C (194 °F)), and followed by one hold at the end for final product extension or brief storage. The temperatures used and the length of time they are applied in each cycle depend on a variety of parameters, including the enzyme used for DNA synthesis, the concentration of bivalent ions and dNTPs in the reaction, and the melting temperature (*Tm*) of the primers. The individual steps common to most PCR methods are as follows:

- Initialization: This step is only required for DNA polymerases that require heat activation by hot-start PCR. It consists of heating the reaction chamber to a temperature of 94–96 °C (201–205 °F), or 98 °C (208 °F) if extremely thermo-stable polymerases are used, which is then held for 1–10 minutes.

- Denaturation: This step is the first regular cycling event and consists of heating the reaction chamber to 94–98 °C (201–208 °F) for 20–30 seconds. This causes DNA melting, or denaturation, of the double-stranded DNA template by breaking the hydrogen bonds between complementary bases, yielding two single-stranded DNA molecules.

- Annealing: In the next step, the reaction temperature is lowered to 50–65 °C (122–149 °F) for 20–40 seconds, allowing annealing of the primers to each of the single-stranded DNA templates. Two different primers are typically included in the reaction mixture: one for each of the two single-stranded complements containing the target region. The primers are single-stranded sequences themselves, but are much shorter than the length of the target region, complementing only very short sequences at the 3' end of each strand.

  It is critical to determine a proper temperature for the annealing step because efficiency and specificity are strongly affected by the annealing temperature. This temperature must be low enough to allow for hybridization of the primer to the strand, but high enough for the hybridization to be specific, i.e., the primer should bind *only* to a perfectly complementary part of the strand, and nowhere else. If the temperature is too low, the primer may bind imperfectly. If it is too high, the primer may not bind at all. A typical annealing temperature is about 3–5 °C below the *Tm* of the primers used. Stable hydrogen bonds between complementary bases are formed only when the primer sequence very closely matches the template sequence. During this step, the polymerase binds to the primer-template hybrid and begins DNA formation.

- Extension/elongation: The temperature at this step depends on the DNA polymerase used; the optimum activity temperature for Taq polymerase is approximately 75–80 °C (167–176 °F), though a temperature of 72 °C (162 °F) is commonly used with this enzyme. In this step, the DNA polymerase synthesizes a new DNA strand complementary to the DNA template strand by adding free dNTPs from the reaction mixture that are complementary to the template in the 5'-to-3' direction, condensing the 5'-phosphate group of the dNTPs with the 3'-hydroxy group at the end of the nascent (elongating) DNA strand. The precise time required for elongation depends both on the DNA polymerase used and on the length of the DNA target region to amplify. As a general rule-of-thumb, at their optimal temperature, most DNA polymerases polymerize a thousand bases per minute. Under optimal conditions (i.e., if there are no limitations due to limiting substrates or reagents), at each extension/elongation step, the number of DNA target sequences is doubled. With each successive cycle, the original template strands plus all newly generated strands become template strands for the next round of elongation, leading to exponential (geometric) amplification of the specific DNA target region.

The processes of denaturation, annealing and elongation constitute a single cycle. Multiple cycles are required to amplify the DNA target to millions of copies. The formula used to calculate the number of DNA copies formed after a given number of cycles is 2^n, where n is the number of cycles. Thus, a reaction set for 30 cycles results in 2^{30}, or 1073741824, copies of the original double-stranded DNA target region.

- Final elongation: This single step is optional, but is performed at a temperature of 70–74 °C (158–165 °F) (the temperature range required for optimal activity of most polymerases used in PCR) for 5–15 minutes after the last PCR cycle to ensure that any remaining single-stranded DNA is fully elongated.

- Final hold: The final step cools the reaction chamber to 4–15 °C (39–59 °F) for an indefinite time, and may be employed for short-term storage of the PCR products.

Ethidium bromide-stained PCR products after gel electrophoresis. Two sets of primers were used to amplify a target sequence from three different tissue samples. No amplification is present in sample #1; DNA bands in sample #2 and #3 indicate successful amplification of the target sequence. The gel also shows a positive control, and a DNA ladder containing DNA fragments of defined length for sizing the bands in the experimental PCRs.

To check whether the PCR successfully generated the anticipated DNA target region (also sometimes referred to as the amplimer or amplicon), agarose gel electrophoresis may be employed for size separation of the PCR products. The size(s) of PCR products is determined by comparison with a DNA ladder, a molecular weight marker which contains DNA fragments of known size run on the gel alongside the PCR products.

Stages

As with other chemical reactions, the reaction rate and efficiency of PCR are affected by limiting factors. Thus, the entire PCR process can further be divided into three stages based on reaction progress:

- *Exponential amplification*: At every cycle, the amount of product is doubled (assuming 100% reaction efficiency). The reaction is very sensitive: only minute quantities of DNA must be present.

- *Leveling off stage*: The reaction slows as the DNA polymerase loses activity and as consumption of reagents such as dNTPs and primers causes them to become limiting.

- *Plateau*: No more product accumulates due to exhaustion of reagents and enzyme.

Optimization

In practice, PCR can fail for various reasons, in part due to its sensitivity to contamination causing amplification of spurious DNA products. Because of this, a number of techniques and procedures have been developed for optimizing PCR conditions. Contamination with extraneous DNA is addressed with lab protocols and procedures that separate pre-PCR mixtures from potential DNA contaminants. This usually involves spatial separation of PCR-setup areas from areas for analysis or purification of PCR products, use of disposable plasticware, and thoroughly cleaning the work surface between reaction setups. Primer-design techniques are important in improving PCR product yield and in avoiding the formation of spurious products, and the usage of alternate buffer components or polymerase enzymes can help with amplification of long or otherwise problematic regions of DNA. Addition of reagents, such as formamide, in buffer systems may increase the specificity and yield of PCR. Computer simulations of theoretical PCR results (Electronic PCR) may be performed to assist in primer design.

Applications

Selective DNA Isolation

PCR allows isolation of DNA fragments from genomic DNA by selective amplification of a specific region of DNA. This use of PCR augments many methods, such as generating hybridization probes for Southern or northern hybridization and DNA cloning, which require larger amounts of DNA, representing a specific DNA region. PCR supplies these techniques with high amounts of pure DNA, enabling analysis of DNA samples even from very small amounts of starting material.

Other applications of PCR include DNA sequencing to determine unknown PCR-amplified sequences in which one of the amplification primers may be used in Sanger sequencing, isolation of a DNA sequence to expedite recombinant DNA technologies involving the insertion of a DNA sequence into a plasmid, phage, or cosmid (depending on size) or the genetic material of another organism. Bacterial colonies (such as E. coli) can be rapidly screened by PCR for correct DNA vector constructs. PCR may also be used for genetic fingerprinting; a forensic technique used to identify a person or organism by comparing experimental DNAs through different PCR-based methods.

Some PCR 'fingerprints' methods have high discriminative power and can be used to identify genetic relationships between individuals, such as parent-child or between siblings, and are used in paternity testing (Fig.). This technique may also be used to determine evolutionary relationships among organisms when certain molecular clocks are used (i.e., the 16S rRNA and recA genes of microorganisms).

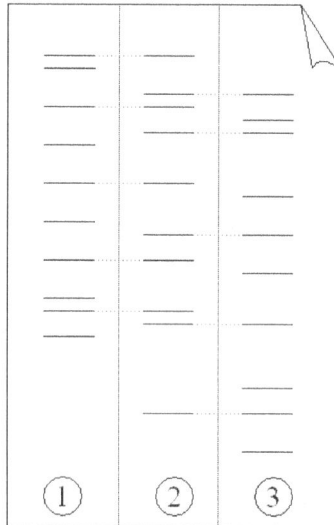

Electrophoresis of PCR-amplified DNA fragments. (1) Father. (2) Child. (3) Mother. The child has inherited some, but not all of the fingerprint of each of its parents, giving it a new, unique fingerprint.

Amplification and Quantification of DNA

Because PCR amplifies the regions of DNA that it targets, PCR can be used to analyze extremely small amounts of sample. This is often critical for forensic analysis, when only a trace amount of DNA is available as evidence. PCR may also be used in the analysis of ancient DNA that is tens of thousands of years old. These PCR-based techniques have been successfully used on animals, such as a forty-thousand-year-old mammoth, and also on human DNA, in applications ranging from the analysis of Egyptian mummies to the identification of a Russiantsar and the body of English king Richard III.

Quantitative PCR methods allow the estimation of the amount of a given sequence present in a sample—a technique often applied to quantitatively determine levels of gene expression. Quantitative PCR is an established tool for DNA quantification that measures the accumulation of DNA product after each round of PCR amplification.

Medical Applications

PCR has been applied to a large number of medical procedures:

- The first application of PCR was for genetic testing, where a sample of DNA is analyzed for the presence of genetic diseasemutations. Prospective parents can be tested for being genetic carriers, or their children might be tested for actually being affected by a disease. DNA samples for prenatal testing can be obtained by amniocentesis, chorionic villus sampling, or even by the analysis of rare fetal cells circulating in the mother's bloodstream. PCR analysis is also essential to preimplantation genetic diagnosis, where individual cells of a developing embryo are tested for mutations.

- PCR can also be used as part of a sensitive test for tissue typing, vital to organ transplantation. As of 2008, there is even a proposal to replace the traditional antibody-based tests for blood type with PCR-based tests.

- Many forms of cancer involve alterations to oncogenes. By using PCR-based tests to study these mutations, therapy regimens can sometimes be individually customized to a patient. PCR permits early diagnosis of malignant diseases such as leukemia and lymphomas, which is currently the highest-developed in cancer research and is already being used routinely. PCR assays can be performed directly on genomic DNA samples to detect translocation-specific malignant cells at a sensitivity that is at least 10,000 fold higher than that of other methods.

Infectious Disease Applications

PCR allows for rapid and highly specific diagnosis of infectious diseases, including those caused by bacteria or viruses. PCR also permits identification of non-cultivatable or slow-growing microorganisms such as mycobacteria, anaerobic bacteria, or viruses from tissue culture assays and animal models. The basis for PCR diagnostic applications in microbiology is the detection of infectious agents and the discrimination of non-pathogenic from pathogenic strains by virtue of specific genes.

Characterization and detection of infectious disease organisms have been revolutionized by PCR in the following ways:

- The human immunodeficiency virus (or HIV), is a difficult target to find and eradicate. The earliest tests for infection relied on the presence of antibodies to the virus circulating in the bloodstream. However, antibodies don't appear until many weeks after infection, maternal antibodies mask the infection of a newborn, and therapeutic agents to fight the infection don't affect the antibodies. PCR tests have been developed that can detect as little as one viral genome among the DNA of over 50,000 host cells. Infections can be detected earlier, donated blood can be screened directly for the virus, newborns can be immediately tested for infection, and the effects of antiviral treatments can be quantified.

- Some disease organisms, such as that for tuberculosis, are difficult to sample from patients and slow to be grown in the laboratory. PCR-based tests have allowed detection of small numbers of disease organisms (both live or dead), in convenient samples. Detailed genetic analysis can also be used to detect antibiotic resistance, allowing immediate and effective therapy. The effects of therapy can also be immediately evaluated.

- The spread of a disease organism through populations of domestic or wild animals can be monitored by PCR testing. In many cases, the appearance of new virulent sub-types can be detected and monitored. The sub-types of an organism that were responsible for earlier epidemics can also be determined by PCR analysis.

- Viral DNA can be detected by PCR. The primers used must be specific to the targeted sequences in the DNA of a virus, and PCR can be used for diagnostic analyses or DNA sequencing of the viral genome. The high sensitivity of PCR permits virus detection soon after infection and even before the onset of disease. Such early detection may give physicians a significant lead time in treatment. The amount of virus ("viral load") in a patient can also be quantified by PCR-based DNA quantitation techniques.

Forensic Applications

The development of PCR-based genetic (or DNA) fingerprinting protocols has seen widespread application in forensics:

- In its most discriminating form, genetic fingerprinting can uniquely discriminate any one person from the entire population of the world. Minute samples of DNA can be isolated from a crime scene, and compared to that from suspects, or from a DNA database of earlier evidence or convicts. Simpler versions of these tests are often used to rapidly rule out suspects during a criminal investigation. Evidence from decades-old crimes can be tested, confirming or exonerating the people originally convicted.

- Less discriminating forms of DNA fingerprinting can help in DNA paternity testing, where an individual is matched with their close relatives. DNA from unidentified human remains can be tested, and compared with that from possible parents, siblings, or children. Similar testing can be used to confirm the biological parents of an adopted (or kidnapped) child. The actual biological father of a newborn can also be confirmed (or ruled out).

Research Applications

PCR has been applied to many areas of research in molecular genetics:

- PCR allows rapid production of short pieces of DNA, even when not more than the sequence of the two primers is known. This ability of PCR augments many methods, such as generating hybridizationprobes for Southern or northern blot hybridization. PCR supplies these techniques with large amounts of pure DNA, sometimes as a single strand, enabling analysis even from very small amounts of starting material.

- The task of DNA sequencing can also be assisted by PCR. Known segments of DNA can easily be produced from a patient with a genetic disease mutation. Modifications to the amplification technique can extract segments from a completely unknown genome, or can generate just a single strand of an area of interest.

- PCR has numerous applications to the more traditional process of DNA cloning. It can extract segments for insertion into a vector from a larger genome, which may be only available in small quantities. Using a single set of 'vector primers', it can also analyze or extract fragments that have already been inserted into vectors. Some alterations to the PCR protocol can generate mutations (general or site-directed) of an inserted fragment.

- Sequence-tagged sites is a process where PCR is used as an indicator that a particular segment of a genome is present in a particular clone. The Human Genome Project found this application vital to mapping the cosmid clones they were sequencing, and to coordinating the results from different laboratories.

- An exciting application of PCR is the phylogenic analysis of DNA from ancient sources, such as that found in the recovered bones of Neanderthals, or from frozen tissues of mammoths. In some cases the highly degraded DNA from these sources might be reassembled during the early stages of amplification.

- A common application of PCR is the study of patterns of gene expression. Tissues (or even individual cells) can be analyzed at different stages to see which genes have become active, or which have been switched off. This application can also use quantitative PCR to quantitate the actual levels of expression

- The ability of PCR to simultaneously amplify several loci from individual sperm has greatly enhanced the more traditional task of genetic mapping by studying chromosomal crossovers after meiosis. Rare crossover events between very close loci have been directly observed by analyzing thousands of individual sperms. Similarly, unusual deletions, insertions, translocations, or inversions can be analyzed, all without having to wait (or pay for) the long and laborious processes of fertilization, embryogenesis, etc.

Limitations

One major limitation of PCR is that prior information about the target sequence is necessary in order to generate the primers that will allow its selective amplification. This means that, typically, PCR users must know the precise sequence(s) upstream of the target region on each of the two single-stranded templates in order to ensure that the DNA polymerase properly binds to the primer-template hybrids and subsequently generates the entire target region during DNA synthesis.

Like all enzymes, DNA polymerases are also prone to error, which in turn causes mutations in the PCR fragments that are generated. Additionally, the specificity of the PCR fragments can mutate to the template DNA, due to nonspecific binding of primers.

Variations

- *Allele-specific PCR*: a diagnostic or cloning technique based on single-nucleotide variations (single-base differences in a patient). It requires prior knowledge of a DNA sequence, including differences between alleles, and uses primers whose 3' ends encompass the SNV (base pair buffer around SNV usually incorporated). PCR amplification under stringent conditions is much less efficient in the presence of a mismatch between template and primer, so successful amplification with an SNP-specific primer signals presence of the specific SNP in a sequence.

- *Assembly PCR* or *Polymerase Cycling Assembly (PCA)*: artificial synthesis of long DNA sequences by performing PCR on a pool of long oligonucleotides with short overlapping segments. The oligonucleotides alternate between sense and antisense directions, and the overlapping segments determine the order of the PCR fragments, thereby selectively producing the final long DNA product.

- *Asymmetric PCR*: preferentially amplifies one DNA strand in a double-stranded DNA template. It is used in sequencing and hybridization probing where amplification of only one of the two complementary strands is required. PCR is carried out as usual, but with a great excess of the primer for the strand targeted for amplification. Because of the slow (arithmetic) amplification later in the reaction after the limiting primer has been used up, extra cycles of PCR are required. A recent modification on this process, known as *Linear-After-The-Exponential-PCR* (LATE-PCR), uses a limiting primer with a higher melting temperature (Tm) than the excess primer to maintain reaction efficiency as the limiting primer concentration decreases mid-reaction.

- *Convective PCR*: a pseudo-isothermal way of performing PCR. Instead of repeatedly heating and cooling the PCR mixture, the solution is subjected to a thermal gradient. The resulting thermal instability driven convective flow automatically shuffles the PCR reagents from the hot and cold regions repeatedly enabling PCR. Parameters such as thermal boundary conditions and geometry of the PCR enclosure can be optimized to yield robust and rapid PCR by harnessing the emergence of chaotic flow fields. Such convective flow PCR setup significantly reduces device power requirement and operation time.

- *Dial-out PCR*: a highly parallel method for retrieving accurate DNA molecules for gene synthesis. A complex library of DNA molecules is modified with unique flanking tags before massively parallel sequencing. Tag-directed primers then enable the retrieval of molecules with desired sequences by PCR.

- *Digital PCR (dPCR)*: used to measure the quantity of a target DNA sequence in a DNA sample. The DNA sample is highly diluted so that after running many PCRs in parallel, some of them do not receive a single molecule of the target DNA. The target DNA concentration is calculated using the proportion of negative outcomes. Hence the name 'digital PCR'.

- *Helicase-dependent amplification*: similar to traditional PCR, but uses a constant temperature rather than cycling through denaturation and annealing/extension cycles. DNA helicase, an enzyme that unwinds DNA, is used in place of thermal denaturation.

- *Hot start PCR*: a technique that reduces non-specific amplification during the initial set up stages of the PCR. It may be performed manually by heating the reaction components to the denaturation temperature (e.g., 95 °C) before adding the polymerase. Specialized enzyme systems have been developed that inhibit the polymerase's activity at ambient temperature, either by the binding of an antibody or by the presence of covalently bound inhibitors that dissociate only after a high-temperature activation step. Hot-start/cold-finish PCR is achieved with new hybrid polymerases that are inactive at ambient temperature and are instantly activated at elongation temperature.

- *In silico PCR* (digital PCR, virtual PCR, electronic PCR, e-PCR) refers to computational tools used to calculate theoretical polymerase chain reaction results using a given set of primers (probes) to amplify DNA sequences from a sequenced genome or transcriptome. In silico PCR was proposed as an educational tool for molecular biology.

- *Intersequence-specific PCR* (ISSR): a PCR method for DNA fingerprinting that amplifies regions between simple sequence repeats to produce a unique fingerprint of amplified fragment lengths.

- *Inverse PCR*: is commonly used to identify the flanking sequences around genomic inserts. It involves a series of DNA digestions and self ligation, resulting in known sequences at either end of the unknown sequence.

- *Ligation-mediated PCR*: uses small DNA linkers ligated to the DNA of interest and multiple primers annealing to the DNA linkers; it has been used for DNA sequencing, genome walking, and DNA footprinting.

- *Methylation-specific PCR* (MSP): developed by Stephen Baylin and Jim Herman at the Johns Hopkins School of Medicine, and is used to detect methylation of CpG islands in genomic DNA. DNA is first treated with sodium bisulfite, which converts unmethylated cytosine bases to uracil, which is recognized by PCR primers as thymine. Two PCRs are then carried out on the modified DNA, using primer sets identical except at any CpG islands within the primer sequences. At these points, one primer set recognizes DNA with cytosines to amplify methylated DNA, and one set recognizes DNA with uracil or thymine to amplify unmethylated DNA. MSP using qPCR can also be performed to obtain quantitative rather than qualitative information about methylation.

- *Miniprimer PCR*: uses a thermostable polymerase (S-Tbr) that can extend from short primers ("smalligos") as short as 9 or 10 nucleotides. This method permits PCR targeting to smaller primer binding regions, and is used to amplify conserved DNA sequences, such as the 16S (or eukaryotic 18S) rRNA gene.

- *Multiplex ligation-dependent probe amplification (MLPA)*: permits amplifying multiple targets with a single primer pair, thus avoiding the resolution limitations of multiplex PCR.

- *Multiplex-PCR*: consists of multiple primer sets within a single PCR mixture to produce amplicons of varying sizes that are specific to different DNA sequences. By targeting multiple genes at once, additional information may be gained from a single test-run that otherwise would require several times the reagents and more time to perform. Annealing temperatures for each of the primer sets must be optimized to work correctly within a single reaction, and amplicon sizes. That is, their base pair length should be different enough to form distinct bands when visualized by gel electrophoresis.

- *Nanoparticle-Assisted PCR (nanoPCR)*: In recent years, it has been reported that some nanoparticles (NPs) can enhance the efficiency of PCR (thus being called nanoPCR), and some even perform better than the original PCR enhancers. It was also found that quantum dots (QDs) can improve PCR specificity and efficiency. Single-walled carbon nanotubes (SWCNTs) and multi-walled carbon nanotubes (MWCNTs) are efficient in enhancing the amplification of long PCR. Carbon nanopowder (CNP) was reported be able to improve the efficiency of repeated PCR and long PCR. ZnO, TiO2, and Ag NPs were also found to increase PCR yield. Importantly, already known data has indicated that non-metallic NPs retained acceptable amplification fidelity. Given that many NPs are capable of enhancing PCR efficiency, it is clear that there is likely to be great potential for nanoPCR technology improvements and product development.

- *Nested PCR*: increases the specificity of DNA amplification, by reducing background due to non-specific amplification of DNA. Two sets of primers are used in two successive PCRs. In the first reaction, one pair of primers is used to generate DNA products, which besides the intended target, may still consist of non-specifically amplified DNA fragments. The product(s) are then used in a second PCR with a set of primers whose binding sites are completely or partially different from and located 3' of each of the primers used in the first reaction. Nested PCR is often more successful in specifically amplifying long DNA fragments than conventional PCR, but it requires more detailed knowledge of the target sequences.

- *Overlap-extension PCR* or *Splicing by overlap extension (SOEing)* : a genetic engineering technique that is used to splice together two or more DNA fragments that contain complementary sequences. It is used to join DNA pieces containing genes, regulatory sequences, or mutations; the technique enables creation of specific and long DNA constructs. It can also introduce deletions, insertions or point mutations into a DNA sequence.

- *PAN-AC*: uses isothermal conditions for amplification, and may be used in living cells.

- *quantitative PCR* (qPCR): used to measure the quantity of a target sequence (commonly in real-time). It quantitatively measures starting amounts of DNA, cDNA, or RNA. quantitative PCR is commonly used to determine whether a DNA sequence is present in a sample and the number of its copies in the sample. *Quantitative PCR* has a very high degree of precision. Quantitative PCR methods use fluorescent dyes, such as Sybr Green, EvaGreen or fluorophore-containing DNA probes, such as TaqMan, to measure the amount of amplified product in real time. It is also sometimes abbreviated to RT-PCR (*real-time* PCR) but this abbreviation should be used only for reverse transcription PCR. qPCR is the appropriate contractions for quantitative PCR (real-time PCR).

- *Reverse Transcription PCR (RT-PCR)*: for amplifying DNA from RNA. Reverse transcriptase reverse transcribes RNA into cDNA, which is then amplified by PCR. RT-PCR is widely used in expression profiling, to determine the expression of a gene or to identify the sequence of an RNA transcript, including transcription start and termination sites. If the genomic DNA sequence of a gene is known, RT-PCR can be used to map the location of exons and introns in the gene. The 5' end of a gene (corresponding to the transcription start site) is typically identified by RACE-PCR (*Rapid Amplification of cDNA Ends*).

- *Solid Phase PCR*: encompasses multiple meanings, including Polony Amplification (where PCR colonies are derived in a gel matrix, for example), Bridge PCR (primers are covalently linked to a solid-support surface), conventional Solid Phase PCR (where Asymmetric PCR is applied in the presence of solid support bearing primer with sequence matching one of the aqueous primers) and Enhanced Solid Phase PCR (where conventional Solid Phase PCR can be improved by employing high Tm and nested solid support primer with optional application of a thermal 'step' to favour solid support priming).

- *Suicide PCR*: typically used in paleogenetics or other studies where avoiding false positives and ensuring the specificity of the amplified fragment is the highest priority. It was originally described in a study to verify the presence of the microbe Yersinia pestis in dental samples obtained from 14th Century graves of people supposedly killed by plague during the medieval Black Death epidemic. The method prescribes the use of any primer combination only once in a PCR (hence the term "suicide"), which should never have been used in any positive control PCR reaction, and the primers should always target a genomic region never amplified before in the lab using this or any other set of primers. This ensures that no contaminating DNA from previous PCR reactions is present in the lab, which could otherwise generate false positives.

- *Thermal asymmetric interlaced PCR (TAIL-PCR)*: for isolation of an unknown sequence flanking a known sequence. Within the known sequence, TAIL-PCR uses a nested pair of primers with differing annealing temperatures; a degenerate primer is used to amplify in the other direction from the unknown sequence.

- *Touchdown PCR (Step-down PCR)*: a variant of PCR that aims to reduce non-specific background by gradually lowering the annealing temperature as PCR cycling progresses. The annealing temperature at the initial cycles is usually a few degrees (3–5 °C) above the T_m of the primers used, while at the later cycles, it is a few degrees (3–5 °C) below the primer T_m. The higher temperatures give greater specificity for primer binding, and the lower temperatures permit more efficient amplification from the specific products formed during the initial cycles.

- *Universal Fast Walking*: for genome walking and genetic fingerprinting using a more specific 'two-sided' PCR than conventional 'one-sided' approaches (using only one gene-specific primer and one general primer—which can lead to artefactual 'noise') by virtue of a mechanism involving lariat structure formation. Streamlined derivatives of UFW are LaNe RAGE (lariat-dependent nested PCR for rapid amplification of genomic DNA ends), 5'RACE LaNe and 3'RACE LaNe.

History

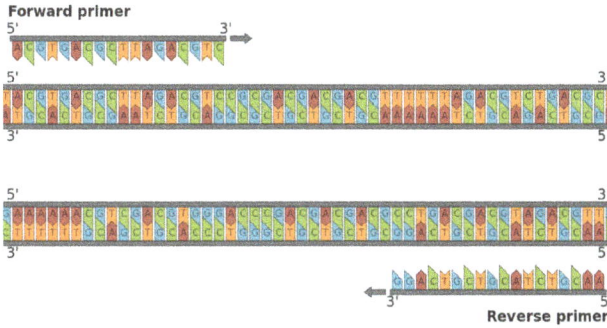

Diagrammatic representation of an example primer pair. The use of primers in an in vitro assay to allow DNA synthesis was a major innovation that allowed the development of PCR.

In the Journal of Molecular Biology by Kjell Kleppe (no) and co-workers in the laboratory of H. Gobind Khorana first described a method using an enzymatic assay to replicate a short DNA template with primers *in vitro*. However, this early manifestation of the basic PCR principle did not receive much attention at the time, and the invention of the polymerase chain reaction in 1983 is generally credited to Kary Mullis.

"Baby Blue", a 1986 prototype machine for doing PCR

When Mullis developed the PCR in 1983, he was working in Emeryville, California for Cetus Corporation, one of the first biotechnology companies. There, he was responsible for synthesizing short chains of DNA. Mullis has written that he conceived of PCR while cruising along the Pacific Coast Highway one night in his car. He was playing in his mind with a new way of analyzing changes (mutations) in DNA when he realized that he had instead invented a method of amplifying any DNA region through repeated cycles of duplication driven by DNA polymerase. In *Scientific American*, Mullis summarized the procedure: "Beginning with a single molecule of the genetic material DNA, the PCR can generate 100 billion similar molecules in an afternoon. The reaction is easy to execute. It requires no more than a test tube, a few simple reagents, and a source of heat." He was awarded the Nobel Prize in Chemistry in 1993 for his invention, seven years after he and his colleagues at Cetus first put his proposal to practice. However, some controversies have remained about the intellectual and practical contributions of other scientists to Mullis' work, and whether he had been the sole inventor of the PCR principle.

At the core of the PCR method is the use of a suitable DNA polymerase able to withstand the high temperatures of >90 °C (194 °F) required for separation of the two DNA strands in the DNA double helix after each replication cycle. The DNA polymerases initially employed for in vitro experiments presaging PCR were unable to withstand these high temperatures. So the early procedures for DNA replication were very inefficient and time-consuming, and required large amounts of DNA polymerase and continuous handling throughout the process.

The discovery in 1976 of Taq polymerase — a DNA polymerase purified from the thermophilic bacterium, *Thermus aquaticus*, which naturally lives in hot (50 to 80 °C (122 to 176 °F)) environments such as hot springs — paved the way for dramatic improvements of the PCR method. The DNA polymerase isolated from *T. aquaticus* is stable at high temperatures remaining active even after DNA denaturation, thus obviating the need to add new DNA polymerase after each cycle. This allowed an automated thermocycler-based process for DNA amplification.

Patent Disputes

The PCR technique was patented by Kary Mullis and assigned to Cetus Corporation, where Mullis worked when he invented the technique in 1983. The *Taq* polymerase enzyme was also covered by patents. There have been several high-profile lawsuits related to the technique, including an unsuccessful lawsuit brought by DuPont. The pharmaceutical company Hoffmann-La Roche purchased the rights to the patents in 1992 and currently holds those that are still protected.

A related patent battle over the Taq polymerase enzyme is still ongoing in several jurisdictions around the world between Roche and Promega. The legal arguments have extended beyond the lives of the original PCR and Taq polymerase patents, which expired on March 28, 2005.

Components Required for PCR Reaction

Following components are required for the PCR reaction

1. DNA template

Those DNA required for amplification that contains the target sequence. At the start of the reaction, high temperature is applied to the original double-stranded DNA molecule to separate the strands from each other (denaturation).

2. DNA Taq polymerase

A type of enzyme that synthesizes new strands of DNA complementary to the target sequence. The first and most commonly used polymerase enzymes is *Taq DNA polymerase* (from *Thermus aquaticus*), whereas *Pfu* DNA polymerase (from *Pyrococcus furiosus*) is used widely because of its higher fidelity when copying DNA. Although these enzymes are subtly different, they both have two capabilities that make them suitable for PCR: (i) they can generate new strands of DNA using a DNA template and primers, and (ii) they are heat resistant.

3. Primers

Short pieces of single-stranded DNA that is complementary to the target sequence. The polymerase begins synthesizing new DNA molecules from the end of the primer.

4. Nucleotides (dNTPs or deoxynucleotide triphosphates)

Single units of the nitrogenous bases A, T, G, and C are required. These are the fundamentally "building blocks" for new DNA strands.

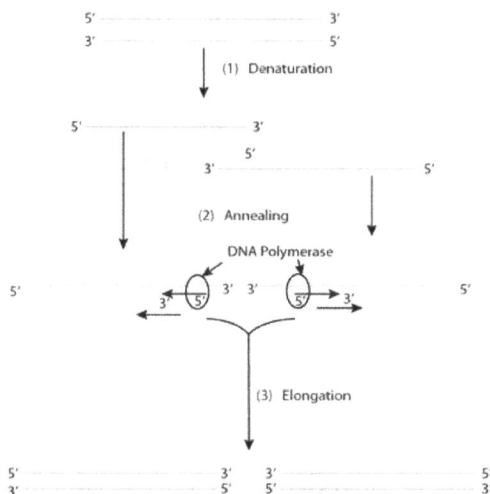

Schematic representation of PCR cycle

Southern Blotting

Southern blotting is one of the central techniques in molecular biology. First devised by E. M. Southern. Southern blotting is the technique used to detect DNA fragments in an agarose gel that are complementary to a given DNA probe. In this method the agarose gel is mounted on a filter-paper wick which dips into a reservoir containing transfer buffer. The hybridization membrane is sandwiched between the gel and a stack of paper towels which serves to draw the transfer buffer through the gel by capillary action. The DNA molecules are carried on the membrane. Nitrocellulose or nylon is used as the membrane material. Gel pretreatment is carried out for efficient Southern blotting.

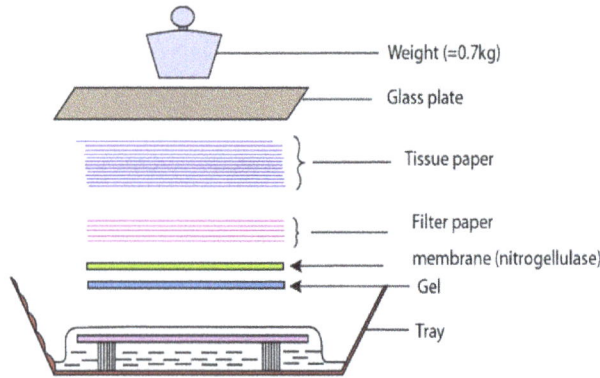

Southern Blotting apparatus

Large DNA fragments require a longer transfer time than short fragments. The electro-phoresed DNA is exposed to a short depurination treatment (0.25 mol/l HCl) followed by alkali for the uniform transfer of a wide range of DNA fragment sizes. This shortens the DNA fragments by alkaline hydrolysis at depurinated sites. It also denatures the fragments prior to transfer allowing them accessible for probing in the single-stranded state. Finally, the gel is equilibrated in neutralizing solution prior to blotting. Then the DNA is transferred in native (non-denatured) form and then alkali-denatured *in situ* on the membrane.

After transfer, the nucleic acid needs to be fixed to the membrane. Oven baking at 80°C is the recommended method for nitrocellulose membranes and this can also be used with nylon membrane. An alternative fixation method utilizes ultraviolet cross-linking. It is based on the formation of cross-links between small fractions of the thymine res-idues in the DNA and positively charged amino groups on the surface of nylon mem-branes. After fixing the nucleic acid, the membrane is placed in a solution of labelled RNA, single-stranded DNA or oligodeoxynucleotide which is complementary in se-quence to the blot transferred DNA bands to be detected. Conditions are chosen so that the labeled nucleic acid hybridizes with the DNA on the membrane. This labeled nucleic acid which is used to detect and locate the complementary sequence is called as the probe .

After the hybridization, the membrane is washed to remove unbound radioactivity and regions of hybridization are detected by using autoradiography. A common ap-proach is to carry out the hybridization under conditions of relatively low stringency which permit a high rate of hybridization, followed by a series of post-hybridization washes of increasing stringency (i.e. higher temperature or lower ionic strength). Autoradiography following each washing stage will reveal any DNA bands that are complementary with the probe and also permit an estimate of the degree of mis-matching to be made. Southern blotting technique is described diagrammatically in figure.

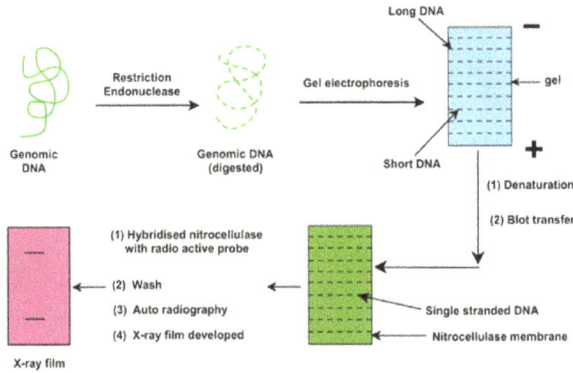

Southern blotting

Applications of Southern Blotting

The Southern blotting methodology can be extremely sensitive. It has a wide range of applications.

1. Important for the confirmation of DNA cloning results.

2. Used for mapping restriction sites around a single copy gene sequence.

3. Detection of minute quantities of DNA in forensic application.

4. Determination of restriction fragment length polymorphism in pathological condition.

5. It is used to find the relation between DNA molecules between species.

Northern Blotting

Northern blotting is the technique used for the specific identification of RNA molecules. This method is similar to southern blotting. But in this case RNAs separated by gel electrophoresis was found not to bind to nitrocellulose. In this procedure the RNA bands are blot-transferred from the gel on to chemically reactive paper, where they are bound covalently. The reactive paper is prepared by diazotization of aminobenzyloxymethyl paper (creating diazobenzyloxymethyl (DBM) paper), which itself can be prepared from Whatman 540 paper by a series of uncomplicated reactions. Later it was found that RNA bands can indeed be blotted on to nitrocellulose membranes under appropriate conditions and suitable nylon membranes have been developed. Once covalently bound, the RNA is available for hybridization with radiolabelled DNA probes. The hybridization can be detected by autoradiography.

Northern blotting is a good technique to determine the number of genes present in a given DNA. The major limitation is that each gene will give rise to two or more RNA transcripts. Another drawback is the presence of exons and introns.

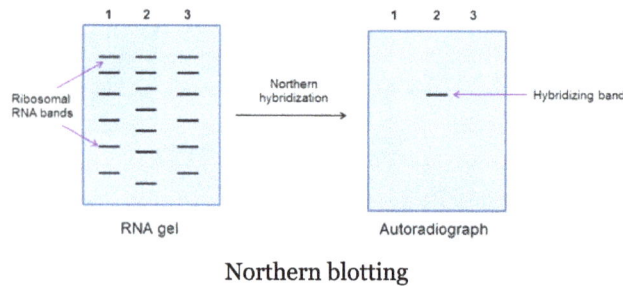

Northern blotting

Dot-blotting

Dot blotting is a modification of southern and northern blotting techniques. In this approach, the nucleic acids are directly spotted onto the filters, and not subjected to electrophoresis. The hybridization procedure is the same as in original blotting techniques. Dot blotting technique is useful in obtaining quantitative data for the evaluation of gene expression.

Western Blotting

Western blotting is a technique used for the identification of proteins. In this procedure there is no direct involvement of nucleic acids, even though it is an important technique for gene manipulation. It involves the transfer of electrophoresed protein bands from a polyacrylamide gel on to a membrane of nitrocellulose or nylon, to which they bind strongly. The bound proteins are then available for analysis by a variety of specific protein–ligand interactions. Most commonly, antibodies are used to detect specific antigens. Lectins have been used to identify glycoproteins.

In these cases the probe may be labeled with radioactivity, or some other 'tag'. In some cases the probe is unlabeled and is itself detected in a 'sandwich' reaction, using a second molecule which is labeled. For example a species-specific second antibody, or protein A of *Staphylococcus aureus* (which binds to certain subclasses of IgG antibodies), or streptavidin (which binds to antibody probes that have been biotinylated) may be labeled in a variety of ways with radioactive, enzyme or fluorescent tags. An advantage of the sandwich approach is that a single preparation of labeled second molecule can be employed as a general detector for different probes. For example, an antiserum may be raised in rabbits, which reacts with a range of mouse immunoglobulins. Such a rabbit anti-mouse (RAM) antiserum may be radiolabeled and used in a number of different applications to identify polypeptide bands probed with different, specific, monoclonal antibodies, each monoclonal antibody being of mouse origin. The sandwich method may also give a substantial increase in sensitivity, owing to the multivalent binding of antibody molecules.

Alternative Blotting Techniques

The original blotting technique employed capillary blotting but nowadays the blotting is usually accomplished by electrophoretic transfer of polypeptides from an SDS-poly-acrylamide gel on to the membrane. Electrophoretic transfer is also the method of choice for transferring DNA or RNA from low-pore-size polyacrylamide gels. It can also be used with agarose gels. However, in this case, the rapid electrophoretic transfer process requires high currents, which can lead to extensive heating effects, resulting in distortion of agarose gels. The use of an external cooling system is necessary to prevent this. Another alternative to capillary blotting is vacuum driven blotting for which several devices are commercially available. Vacuum blotting has several advantages over capillary or electrophoretic transfer methods: transfer is very rapid and gel treatment can be performed *in situ* on the vacuum apparatus. This ensures minimal gel handling and, together with the rapid transfer, prevents significant DNA diffusion.

Gene Silencing

The term gene silencing is commonly used to describe the "switching off" of a gene by a mechanism without genetic modification. The term gene silencing refers to an epigenetic phenomenon, the heritable inactivation of gene expression that does not involve any changes to the deoxyribonucleic acid (DNA) sequence. While this phenomenon has initially been studied in transgenic plants, its relevance in the regulation of endogenous plant genes has become increasingly apparent. Below some cellular components are mentioned where gene silencing occurred:

- Chromatin and heterochromatin
- Dicer
- dsRNA
- Histones
- MicroRNA
- siRNA
- Transposons

Gene silencing has following two major subdivisions by which genes are regulated:

1. Transcriptional gene silencing (TGS) and
2. Posttranscriptional gene silencing (PTGS)

Transcriptional Gene Silencing (TGS)

Transcriptional gene silencing is the product of chromosomal histone modifications, creating an environment of heterochromatin, which is surrounded to a gene that makes it inaccessible to transcriptional machinery (RNA polymerase, transcription factors, etc.). TGS blocks primary transcription from nuclear DNA and is in most cases associated with DNA methylation and chromatin condensation in nearly all organisms that possess a DNA methylation system.

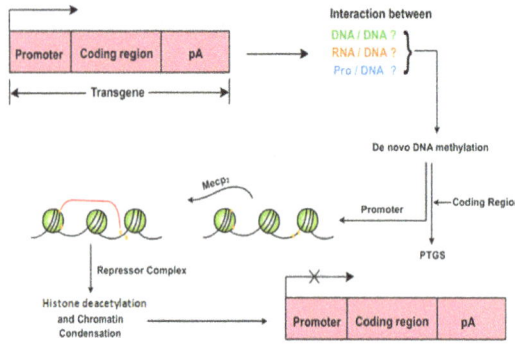

Schematic representation of DNA methylation – mediated transcriptional gene silencing (TGS)

Post-transcriptional Gene Silencing (PTGS)

Post-transcriptional gene silencing is the product of transcribed mRNA of a specific gene being silenced. When mRNA was destructed, then translation to form an active gene product (in most cases, a protein) will be prevented. A general process of post-transcriptional gene silencing is by RNAi. PTGS involves a cytoplasmic, target sequence-specific RNA degradation process that is possibly activated by double-stranded RNA (dsRNA). This dsRNA is independent of ongoing translation.

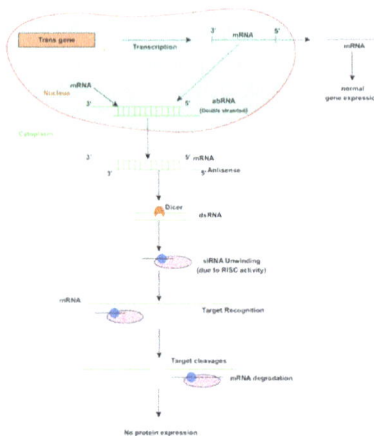

Schematic representation of post translational gene silencing (PTGS)

TGS can be transmitted generation to generation by meiosis whereas PTGS is usually lost during meiosis. In PTGS, double stranded RNA is interred into a cell and gets chopped up by the enzyme known as dicer to form siRNA. siRNA then binds to the RNA-induced silencing complex (RISC) and is unwound. The anitsense RNA complexed with RISC protein and binds to its corresponding mRNA, which is then cleaved by the enzyme slicer rendering it inactive.

Applications of Genetic Engineering

The knowledge of the molecular basis of diseases caused by various pathogens has allowed testing different strategies to produce disease resistant transgenic plants. Genetic engineering has also been successful in producing herbicide resistance plants. Some other applications are to develop high degree of tolerance or resistance to pests (insects, nematodes, etc.) and diseases. Below are some examples of genetic engineering applications in agriculture:

- Virus resistance

- Insect resistance

- Golden rice

- Long lasting tomatoes

Virus Resistance

Plant viruses can cause severe damage to crops by substantially reducing vigor, yield, and product quality. Viruses cause more than 1400 plant diseases and thus, decreasing the agricultural productivity. Unfortunately, there is no viracidal compound to control these diseases. Some diseases, such as rice tungro disease, are caused by two or more distinct viruses and attempts to incorporate genes for resistance against them have not met with success. Virus resistance is achieved usually through the antiviral pathways of RNA silencing, a natural defense mechanism of plants against viruses. The experimental approach consists of isolating a segment of the viral genome itself and transferring it into the genome of a susceptible plant. Integrating a viral gene fragment into a host genome does not cause disease (the entire viral genome is needed to cause disease). Instead, the plant's natural antiviral mechanism that acts against a virus by degrading its genetic material in a nucleotide sequence specific manner via a cascade of events involving numerous proteins, including ribonucleases (enzymes that cleave RNA), is activated. This targeted degradation of the genome of an invader virus protects plants from virus infection.

Three hypothesis have been investigated to engineer development of virus resistance plants:

- Expression of the virus coat protein (CP) gene

- Expression of satellite RNAs and

- Use of antisense viral RNA

Expression of the Virus Coat Protein (CP) Gene

This technique is most common. In this CP-mediated resistance (CP-MR) is developed, based on the well known process of cross protection. It is protected against super infection by a severe strain of related virus. This method has been commonly used in agriculture to confer protection against severe virus infection. However, the technique has following disadvantages:

- Due to synergistic interaction, infection of cross protected plants with a second unrelated virus may cause a severe disease,

- The suspicious virus strain might mutate to a more severe form, leading to extensive crop losses,

- Protecting virus strain may cause a small but significant decrease in yields, and

- In cross protection, the protecting virus must be applied each growing season.

Most of these problems can be offset by genetic engineering of CP-MR in plants. CP-MR produced a c-DNA encoding the capsid protein (CP) sequences of TMV, ligated it to a strong transcriptional promoter (CaMV 35S promoter) and transport sequences to provide constitutive expression of the gene throughout the transgenic plant, and flanked on the 3' end by poly A signal from the nopaline synthase gene. This chimeric gene was introduced into a disarmed plasmid of *A. tumefaciens* and the modified bacterium was used.

Expression of Satellite RNAs

Some viruses have specific feature to contain, in addition to their genomic RNA, a small RNA molecule known as satellite RNA (S-RNA). The S-RNAs require the company of a specific 'helper' virus (closely related virus) for their replication. S-RNA does not have sequence to encode CP. They are encapsulated in the coat protein of their helper virus or satellite vi-ruses which encode their own coat protein. Due to ability to modify disease symptoms S-RNAs now have point of attention in genetic engineering. Most of the S-RNAs decrease the severity of viral infection, presumably through interference with viral replication. By this method, tomato, a number of pepper varieties, cucumber, eggplant, cabbage and tobacco plants against CMV have been protected. The first time S-RNA induced attenuation of viral symptoms involved the introduction of cDNA copies of CMV S-RNA into the genome of tobacco plants.

Use of Antisense Viral RNA

Here using the antisense RNA, which is a single stranded RNA molecule complementary to the mRNA (sense RNA), transcribed by a given gene, is another approach suggested for introducing viral resistance in plants. The sense RNA carries codons to translate to a specified sequence of amino acids. The antisense RNA, on the other hand, does not contain the functional protein sequences. When both sense and antisense RNA are present together in cytoplasm they anneal to form a duplex RNA molecule which cannot be translated. Using this methodology, transgenic plants expressing 3' region of antisense RNA, including CP gene of TMV or CMV. RNAs were produced which have property to protect against infection with respective viruses or viral RNA.

Insect Resistance

Insects cause serious losses in agricultural products in the field at the time of cultivation and during storage. Insects belonging to the orders, Coleoptera, Lepidoptera and Diptera, are the most serious plant pests which cause agricultural damages. Use of insecticides, bio-pesticides has several harmful side effects. *Bacillus thuringiensis* (Bt), a free-living, Gram-positive soil bacterium, has been employed as insecticide specificity towards lepidopteran pests. It is environmentally safe and thus, is high in demand. On the other hand, the major problems in using Bt sprays for controlling the insect attack on plants are:

- The high cost of production of Bt insecticide and

- The instability of the protoxin crystal proteins under field conditions, necessitating multiple applications.

To avoid these problems transgenic plants expressing Bt toxin genes have been engineered. Insect resistant transgenic plants have also been created by introducing trypsin inhibitor gene.

Bt Cotton

Two Bt proteins have been recognized as being of particular use for the control of the major pests of cotton and the genes encoding for these proteins have been incorporated into cotton plants by Monsanto. In the 1980's a lot work was undertaken by Monsanto to identify and extract the Bt genes and during this decade the gene encoding for the Bt protein Cry1Ac was successfully inserted into a cotton plant. Nowadays, several plant genes are transformed and used as insect resistant plants.

Golden Rice

Golden rice is genetically modified rice which contains a large amount of A-vitamins. Or more correctly, the rice contains the constituent beta-carotene which is converted in the body into Vitamin-A. So when you eat golden rice, so can get more amount of vitamin of A. Beta-carotene is orange colour so genetically modified rice is golden color. For the making of golden for synthesis of beta-carotene three new genes are implanted: two from daffodils and the third from a bacterium.

Advantages:

- The rice can be considered for poor people in underdeveloped countries. They eat only an extremely limited diet lacking in the essential bodily vitamins.

Disadvantage

- Critics terror that poor people in underdeveloped countries are becoming too dependent on the rich western world. Generally, genetically modified plants are developed by the large private companies in the West.

- The customers who buy patented transgenic seeds from the company may need to sign a contract not to save or sell the seeds from their harvest, which raises concerns that this technology might lead to dependence for small farmers.

Long-lasting Tomatoes

Long-lasting, genetically modified tomatoes now came in to the market. This is the first genetically modified food available to consumers. The genetically modified tomato produces less of the substance that causes tomatoes to rot, so remains firm and fresh for a long time.

References

- Monroe D. (2006). "Jumping Genes Cross Plant Species Boundaries". PLoS Biology. 4 (1): e35. doi:10.1371/journal.pbio.0040035

- A. Pingoud (2004). Restriction Endonucleases (Nucleic Acids and Molecular Biology). Springer. p. 3. ISBN 9783642188510

- ISAAA 2013 Annual Report Executive Summary, Global Status of Commercialized Biotech/GM Crops: 2013 ISAAA Brief 46-2013, Retrieved 6 August 2014

- Maghari, Behrokh Mohajer, and Ali M. Ardekani. "Genetically Modified Foods And Social Concerns." Avicenna Journal Of Medical Biotechnology 3.3 (2011): 109-117. Academic Search Premier. Web. 7 Nov. 2014

- Russell DW, Sambrook J (2001). Molecular cloning: a laboratory manual. Cold Spring Harbor, N.Y: Cold Spring Harbor Laboratory. ISBN 0-87969-576-5

- ISAAA 2015 Annual Report Executive Summary, 20th Anniversary (1996 to 2015) of the Global Commercialization of Biotech Crops and Biotech Crop Highlights in 2015 ISAAA Brief 51-2015, Retrieved 19 August 2016

- Shrawat, A.; Lörz, H. (2006). "Agrobacterium-mediated transformation of cereals: a promising approach crossing barriers". Plant biotechnology journal. 4 (6): 575–603. doi:10.1111/j.1467-7652.2006.00209.x. PMID 17309731

- Krieger M, Scott MP, Matsudaira PT, Lodish HF, Darnell JE, Zipursky L, Kaiser C, Berk A (2004). Molecular Cell Biology (5th ed.). New York: W.H. Freeman and Company. ISBN 0-7167-4366-3

- Maxmen, Amy (2 May 2012) First plant-made drug on the market Nature, Biology & Biotechnology, Industry. Retrieved 26 June 2012

- Koornneef, M.; Meinke, D. (2010). "The development of Arabidopsis as a model plant". The Plant journal : for cell and molecular biology. 61 (6): 909–921. doi:10.1111/j.1365-313X.2009.04086.x. PMID 20409266

- Alexander N. Glazer; Hiroshi Nikaidō (2007). Microbial biotechnology: fundamentals of applied microbiology. Cambridge University Press. ISBN 0-521-84210-7

- Areal, F. J.; Riesgo, L.; Rodríguez-Cerezo, E. (2012). "Economic and agronomic impact of commercialized GM crops: A meta-analysis". The Journal of Agricultural Science. 151: 7–33. doi:10.1017/S0021859612000111

- Bartlett, J. M. S.; Stirling, D. (2003). "A Short History of the Polymerase Chain Reaction". PCR Protocols. Methods in Molecular Biology. 226 (2nd ed.). pp. 3–6. doi:10.1385/1-59259-384-4:3. ISBN 1-59259-384-4

- Brookes, Graham and Barfoot, Peter (May 2012) GM crops: global socio-economic and environmental impacts 1996-2010 PG Economics Ltd. UK, Retrieved 3 January 2012

- Tennille, Tracy (February 13, 2015). "First Genetically Modified Apple Approved for Sale in U.S.". Wall Street Journal. Retrieved October 3, 2016

- J., Ninfa, Alexander; P., Ballou, David (2004). Fundamental laboratory approaches for biochemistry and biotechnology. Wiley. ISBN 1891786008. OCLC 633862582

- About Golden Rice Archived 2 November 2012 on Wayback Machine. International Rice Research Institute. Retrieved 20 August 2012

- Wang, Guo-ying (2009). "Genetic Engineering for Maize Improvement in China". Electronic Journal of Biotechnology. Electronic Journal of Biotechnology. Retrieved December 1, 2015

- P., Ballou, David; Marilee., Benore, (2010). Fundamental laboratory approaches for biochemistry and biotechnology. John Wiley. ISBN 9780470087664. OCLC 420027217

- "Genetically Engineered Foods - Plant Virus Resistance" (PDF). Cornell Cooperative Extension. Cornell University. 2002. Retrieved October 3, 2016

Permissions

All chapters in this book are published with permission under the Creative Commons Attribution Share Alike License or equivalent. Every chapter published in this book has been scrutinized by our experts. Their significance has been extensively debated. The topics covered herein carry significant information for a comprehensive understanding. They may even be implemented as practical applications or may be referred to as a beginning point for further studies.

We would like to thank the editorial team for lending their expertise to make the book truly unique. They have played a crucial role in the development of this book. Without their invaluable contributions this book wouldn't have been possible. They have made vital efforts to compile up to date information on the varied aspects of this subject to make this book a valuable addition to the collection of many professionals and students.

This book was conceptualized with the vision of imparting up-to-date and integrated information in this field. To ensure the same, a matchless editorial board was set up. Every individual on the board went through rigorous rounds of assessment to prove their worth. After which they invested a large part of their time researching and compiling the most relevant data for our readers.

The editorial board has been involved in producing this book since its inception. They have spent rigorous hours researching and exploring the diverse topics which have resulted in the successful publishing of this book. They have passed on their knowledge of decades through this book. To expedite this challenging task, the publisher supported the team at every step. A small team of assistant editors was also appointed to further simplify the editing procedure and attain best results for the readers.

Apart from the editorial board, the designing team has also invested a significant amount of their time in understanding the subject and creating the most relevant covers. They scrutinized every image to scout for the most suitable representation of the subject and create an appropriate cover for the book.

The publishing team has been an ardent support to the editorial, designing and production team. Their endless efforts to recruit the best for this project, has resulted in the accomplishment of this book. They are a veteran in the field of academics and their pool of knowledge is as vast as their experience in printing. Their expertise and guidance has proved useful at every step. Their uncompromising quality standards have made this book an exceptional effort. Their encouragement from time to time has been an inspiration for everyone.

The publisher and the editorial board hope that this book will prove to be a valuable piece of knowledge for students, practitioners and scholars across the globe.

Index

—